Vertex Algebras and Integral Bases for the Enveloping Algebras of Affine Lie Algebras

Recent Titles in This Series

(*Continued in the back of this publication*)

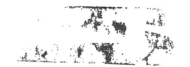

MEMOIRS
of the
American Mathematical Society

Number 466

Vertex Algebras and Integral Bases for the Enveloping Algebras of Affine Lie Algebras

Shari A. Prevost

March 1992 • Volume 96 • Number 466 (end of volume) • ISSN 0065-9266

American Mathematical Society
Providence, Rhode Island

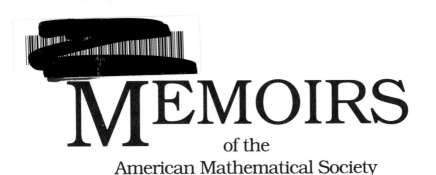

1991 *Mathematics Subject Classification.*
Primary 17B35, 17B65, 17B67.

Library of Congress Cataloging-in-Publication Data

Prevost, Shari A., 1960-
Vertex algebras and integral bases for the enveloping algebras of affine Lie algebras/Shari A. Prevost.
p. cm. - (Memoirs of the American Mathematical Society, ISSN 0065-9266; no. 466)
Revision of the author's thesis (Ph.D.)–Rutgers University, 1989.
"March 1992."
"Volume 96 number 466."
Includes bibliographical references.
ISBN 0-8218-2527-5
1. Universal enveloping algebras. 2. Lie algebras. 3. Kac-Moody algebras. I. Title. II. Series.
QA3.A57 no. 466
[QA252.3]
510 s–dc20 91-44874
[512′.55] CIP

SUBSCRIPTION INFORMATION. The 1992 subscription begins with Number 459 and consists of six mailings, each containing one or more numbers. Subscription prices for 1992 are $292 list, $234 institutional member. A late charge of 10% of the subscription price will be imposed on orders received from nonmembers after January 1 of the subscription year. Subscribers outside the United States and India must pay a postage surcharge of $25; subscribers in India must pay a postage surcharge of $43. Expedited delivery to destinations in North America $30; elsewhere $82. Each number may be ordered separately; *please specify number* when ordering an individual number. For prices and titles of recently released numbers, see the New Publications sections of the NOTICES of the American Mathematical Society.

BACK NUMBER INFORMATION. For back issues see the AMS Catalogue of Publications.

Subscriptions and orders for publications of the American Mathematical Society should be addressed to American Mathematical Society, Box 1571, Annex Station, Providence, RI 02901-1571. *All orders must be accompanied by payment.* Other correspondence should be addressed to Box 6248, Providence, RI 02940-6248.

Memoirs of the American Mathematical Society (ISSN 0065-9266) is published bimonthly (each volume consisting usually of more than one number) by the American Mathematical Society at 201 Charles Street, Providence, Rhode Island 02904-2213. Second Class postage paid at Providence, Rhode Island 02940-6248. Postmaster: Send address changes to Memoirs of the American Mathematical Society, American Mathematical Society, Box 6248, Providence, RI 02940-6248.

Printed in the United States of America.
The paper used in this book is acid-free and falls within the guidelines established to ensure permanence and durability. ♾

10 9 8 7 6 5 4 3 2 1 97 96 95 94 93 92

Table of Contents

Abstract

We present a new proof of the identities needed to exhibit an explicit **Z**-basis for the universal enveloping algebra associated to an affine Lie algebra. We then use the explicit **Z**-bases to extend Borcherds' description, via vertex operator representations, of a **Z**-form of the enveloping algebras for the simply-laced affine Lie algebras to the enveloping algebras associated to the unequal root length affine Lie algebras.

Keywords and phrases. Affine Lie algebra, integral basis, universal enveloping algebra, vertex operator algebra.

Received by the editor August 27, 1990.

Chapter 1
Introduction

The vertex operator made its debut in the mathematics literature a little more than a decade ago. Since then, the theory of vertex operators has flourished. Vertex operator representations have been shown to yield all of the finite-dimensional irreducible representations of the simple Lie algebras. Also vertex operators have provided a very natural setting in which to study affine Lie algebras and their representations. And perhaps the most important contribution, to date, of vertex operator theory is the construction of the Monster and the Moonshine module $V^\natural$ [F-L-M], [B]. It should also be noted that vertex operator theory plays a fundamental role in string theory. The reader may refer to the introduction in [F-L-M] for a discussion of contributions to vertex operator theory by physicists.

This paper utilizes the vertex operator representations of the affine Lie algebras to give two different, yet equivalent, descriptions of integral bases for the affine Lie algebras and their associated universal enveloping algebras. First we present a new proof for the straightening arguments used by Garland in [G] and later by Mitzman in [M], to exhibit an explicit **Z**-form for the universal enveloping algebras associated to the type 1, and type 2 and type 3 Lie algebras. We remark that the identities which allow one to "straighten" the **Z**-basis elements are the most crucial part of these proofs. Secondly, we shift our attention to the vertex operator descriptions of an integral (or **Z**-) form of the universal enveloping algebra of an affine Lie algebra. We first use vertex operator algebra theory to rewrite the explicit **Z**-bases for the enveloping algebras of the affine Lie algebras exhibited by both Garland [G] and Mitzman [M].

Next we examine the vertex algebra approach Borcherds uses in his announcement [B] to give a (nonexplicit) description of an integral form of the universal enveloping algebra of a simply-laced affine Lie algebra. Then we extend the method found in [B] to give an analogous description of a **Z**-form of the remaining unequal root length affines.

A vertex algebra consists of a **Z**-graded vector space $V = \coprod_{n\in\mathbf{Z}} V_n$ together with a linear map $v \mapsto Y(v,z)$ from V into the vector space $(End\, V)[[z,z^{-1}]]$ and two distinguished vectors **1** and ω in V so that a number of properties, such as the Jacobi identity, hold. If in addition to the above properties the vector space V satisfies the conditions that $dim\ V_n < \infty$ for all n and $dim\ V_n = 0$ for sufficiently small n, then V is called a vertex operator algebra. The formal Laurent series $Y(v,z) = \sum_{n\in\mathbf{Z}} v_n z^{-n-1}$ are called vertex operators. Since vertex operator algebras are actually vertex algebras, one could use the term "restricted" vertex algebra for vertex operator algebra. However, we choose to use the two reference terms "vertex algebra" [B] and "vertex operator algebra" [F-L-M] to distinguish the two concepts in our work; both structures are useful. We also note that our definition of vertex algebra and the definition of vertex operator algebra (which is the same as in [F-L-M]) are modifications of Borcherds' definition of a vertex algebra used in [B].

An integral (or **Z**-) form $A_{\mathbf{Z}}$ of an algebra A over a field **F** is an algebra over **Z** such that $A_{\mathbf{Z}} \otimes_{\mathbf{Z}} \mathbf{F} = A$. Integral bases and integral forms for finite-dimensional Lie algebras were first formulated in general by Chevalley [Chev]. These bases, called Chevalley bases, have integral structure constants which are related to root strings and are such that the **Z**-span of the Chevalley bases is invariant under the action of $\frac{(ad\, x_\alpha)^m}{m!}$, where x_α is a root vector contained in the Chevalley basis and m is a nonnegative integer. The latter property led to the construction of Chevalley groups (of adjoint type) ([Chev], cf. [H], [Stein]). Later Kostant [Kost], and also Cartier (cf. [T]), discovered and gave an explicit description of a **Z**-form generated by the elements $\frac{(x_\alpha)^m}{m!}$ for the universal enveloping algebras associated to these finite-dimensional Lie algebras. This played a role in a generalization of the definition of a Chevalley group [Stein] which utilized representations other than the adjoint representation of a Lie algebra (cf. [H]).

To describe a natural integral form for the universal enveloping algebra of an affine Lie algebra, one has a choice of two methods: the explicit construction of a **Z**-basis of the enveloping algebra via a Chevalley-type basis of the affine Lie algebra, as in [G] or [M], or a (nonexplicit) description in which some set of conditions is used to define a **Z**-form of the enveloping algebra. The second method is used in [T] and [B]. Tits in [T] used a certain set of generators to obtain an integral form. For the simply-laced affine Lie algebras, Borcherds in [B] employed the theory of vertex algebras to define a new family of irreducible integrable modules for these affine algebras and to find integral forms for the modules. He then defined the integral form of the universal enveloping algebra to consist of the elements of the enveloping algebra which preserve the integral forms of the modules in the aforementioned family.

In the present work we give a new proof for the identities which allow one to straighten the **Z**-basis elements in the integral form of the universal enveloping algebras. The proof offered here differs in two respects from the previous proof. We first understand how to prove the identities in the enveloping algebras associated to the simply-laced affines. Then we observe that the identities for the remaining affines follow as corollaries to the identities in the equal root length affines. Secondly, the proof we present is constructive in nature: the earlier proofs merely "checked" the validity, but didn't give much insight into the discovery of the identities.

Next we consider the vertex operator representations of the affine Lie algebras. It was discovered in [L-W1], [K-K-L-W], [F-K] and [Seg] that affine Lie algebras could be constructed explicitly using vertex operator representations. See also the introduction in [F-L-M] for a discussion of the contributions to this phenomenom made by physicists.

We also give a detailed version of the method employed by Borcherds in his announcement [B] to find a (nonexplicit) description of an integral form (but not a $\mathbf{Z}$-basis) for the universal enveloping algebra of a simply-laced affine Lie algebra. To show that the $\mathbf{Z}$ subalgebra defined by Borcherds is indeed an integral form, we use the discussion in [L-P] and [Mac] concerning Schur polynomials and a basis of the symmetric algebra $S(\hat{\underline{\mathfrak{h}}}^-_{\mathbf{Z}})$. Schur polynomials also arise in calculations involving the Virasoro operator $L(-1)$. We first demonstrate that the $\mathbf{Z}$-subalgebra defined by Borcherds contains (with a slight alteration) the $\mathbf{Z}$-form generated by the $\mathbf{Z}$-basis given by Garland [G] and Mitzman [M]. (The explicit description is found in Theorem 4.2.6 below.) Then using a standard diagonal map argument, we prove that the (slightly altered) $\mathbf{Z}$-form obtained from the $\mathbf{Z}$-basis in Theorem 4.2.6 must contain the $\mathbf{Z}$-subalgebra defined by Borcherds.

We define a $\mathbf{Z}$-subalgebra of the enveloping algebra of an unequal root length affine which is analogous to the $\mathbf{Z}$-subalgebra described in Borcherds' announcement [B] for the enveloping algebra of a simply-laced affine. We then show this is an integral form of the enveloping algebra, and is (essentially) the same integral form generated by the $\mathbf{Z}$-basis of Theorem 4.2.6.

The material of this paper is organized in the following manner.

In Chapter 2 we briefly rewrite Chapter 2 and Chapter 3 of [M], i.e., the construction of the simply-laced affine algebras and the unequal root length affine algebras. To recover the unequal root length affines, we consider these algebras as subalgebras of the simply-laced affines fixed by a suitable extension $\hat{\nu}$ of a Dynkin diagram automorphism of order two or three (cf. [K-K-L-W], [M]).

In Chapter 3 we give our treatment of the main theorem of [G] and [M]. This theorem (Theorem 3.1.6) gives an explicit description of an integral basis for the universal enveloping algebra for any of the affine Lie algebras discussed in Chapter 2. While the overall idea of the proof of this theorem is the same as that of Theorem 4.2.6 in [M], the proofs of the commutator identities needed for the straightening arguments are all new. Except for Proposition 3.2.2 below, the arguments which prove the identities use techniques different from those employed in [M]: the verifications of the formulas are direct computations using Theorem 3.1.5 and the commutation relations found in Chapter 2. Theorem 3.1.5 presents a new commutation identity involving exponentials of formal power series (with zero constant term) whose coefficients are elements of an affine Lie algebra. Although Proposition 3.2.2 is proved using a method similar to the procedure Mitzman relied on, its statement is new; it is a generalization of Lemma 4.3.4 (iii) in [M] for the simply-laced affines. We note that it is only in this instance (Proposition 3.2.2) that we must resort to such tactics (i.e., a proof which doesn't compute the formula directly)! Our treatment of the formulas for the unequal root length affines differs completely from that of [M] and [G]. We prove the identities in a case by case manner; the cases are divided naturally by the order of the graph automorphism and by the inner products of the different roots. We first derive the identities for the simply-laced affines and then use these formulas (along with Theorem 3.1.5 and the Baker-Campbell-Hausdorff Theorem) to give constructive proofs of the needed identities for the unequal root length affines.

In Sections 4.1 and 4.2 we present a brief account of the background material needed to place the problem of finding an explicit description of a **Z**-form for the universal enveloping algebras of the affine Lie algebras within the realm of vertex operator theory. Section 4.2 introduces the notion of vertex operator algebra as found in [F-L-M]. Here we also define the closely related concept of vertex algebra. Such structures were first discussed in [B], although Borcherds' definition does not include either the Jacobi identity or the properties of the Virasoro algebra. In the same section we also describe the construction of a vertex operator algebra V_L for an even positive definite lattice L. The reader may wish to consult Chapters 5 - 8 of [F-L-M] for a more thorough account. Then Section 4.2 recalls the construction of [F-K], [Seg] of the simply-laced affine Lie algebras using the central extension of thc lattice L utilized in Section 4.1 to construct V_L. We also give a definition of a Chevalley basis and a description of such a basis for the simply-laced affine Lie algebras, as in [G] and [M].

In Sections 4.3 - 4.7 we first present a version of Borcherds' use of vertex algebras to give a (nonexplicit) description of an integral form of the universal enveloping algebras for the simply-laced affine algebras. The relationship between Schur polynomials and a basis for the symmetric algebra $S(\hat{\underline{h}}_{\mathbf{Z}}^{-})$ play an important role in verifying Borcherds' assertion that this description is an integral form. One difference between Borcherds' approach and our approach is that one considers the affine Lie algebra as a subquotient of the vertex algebra instead of looking at the Lie algebra generated by the components of the vertex operators associated with the weight one elements of a vertex operator algebra. Secondly, we show that the integral form of the universal enveloping algebra obtained using the vertex algebra approach is (essentially) (cf. (4.6.3)) the same as the integral form obtained from the subalgebra generated by the **Z**-basis explicitly described in Theorem 4.2.6. Finally, in Section 4.7 we extend the Borcherds-type vertex algebra description of a **Z**-form of the universal enveloping algebra to include the remaining unequal root length affines.

Acknowledgements. This paper is a revised version of the author's Ph.D. dissertation at Rutgers University, 1989. I would like to thank my advisor Professor Robert Wilson and Professor James Lepowsky for all of their invaluable help throughout my stay at Rutgers University, and especially for their time spent reading this work.

Chapter 2

Construction of the affine Lie algebras

2.1 $A_l^{(1)}$ $(l \geq 1)$, $D_l^{(1)}$ $(l \geq 1)$ and $E_l^{(1)}$ $(l = 6,7,8)$

In this section we briefly review the construction of the simply-laced (or equal root length) Lie algebras and their affinizations from a positive definite even lattice L with a symmetric bilinear form $\langle \cdot, \cdot \rangle$. We use the "vocabulary" of Chapter 6 of [F-L-M] to rewrite Chapter 2 of [M]. For a more complete treatment of the above and for other properties of affine Lie algebras, see [B], [F-K], [F-L-M], [G], [Kac], [L], [L-W1], [L-W2], [M], or [Seg].

Let L be an even nondegenerate lattice of rank l $(l \geq 1)$, with a symmetric bilinear form $\langle \cdot, \cdot \rangle : L \times L \to \mathbf{Z}$. Suppose further that L is spanned (over $\mathbf{Z}$) by the set

$$\Delta = \{\alpha \in L \mid \langle \alpha, \alpha \rangle = 2\}, \tag{2.1.1}$$

where Δ is a rank l indecomposable root system of type A_l $(l \geq 1)$, D_l $(l \geq 1)$ or E_l $(l = 6,7,8)$. Let $\Pi = \{\alpha_1, \alpha_2, \ldots, \alpha_l\}$ denote the simple roots of Δ. Note that Π forms a $\mathbf{Z}$-basis of the root system Δ, and hence, also of L.

Let $(\hat{L}, -)$ be a central extension of L by the cyclic group $\langle \kappa \mid \kappa^2 = 1 \rangle$, i.e.,

$$1 \to \left\langle \kappa \mid \kappa^2 = 1 \right\rangle \hookrightarrow \hat{L} \xrightarrow{-} L \to 0 \tag{2.1.2}$$

and let $c_0 : L \times L \to \mathbf{Z}/2\mathbf{Z}$ be the associated commutator map determined by

$$ab = \kappa^{c_0(\bar{a}, \bar{b})} ba \tag{2.1.3}$$

for $a,\ b \in \hat{L}$. Assume that $c_0(\bar{a}, \bar{b}) = \langle \bar{a}, \bar{b} \rangle \bmod 2\mathbf{Z}$. Let $e : L \to \hat{L}$ be a section with corresponding 2-cocycle $\epsilon_0 : L \times L \to \mathbf{Z}/2\mathbf{Z}$ such that

$$e(\alpha) \mapsto e_\alpha$$

$$e(0) \mapsto e_0 \tag{2.1.4}$$

$$e_\alpha e_\beta = e_{\alpha+\beta} \kappa^{\epsilon_0(\alpha,\beta)}$$

for $\alpha,\ \beta \in \Delta$. Let $\hat{\Delta}$ denote the pullback of the root system Δ in $\hat{L}$; thus

$$\hat{\Delta} = \{a \in \hat{L} \mid \bar{a} \in \Delta\}. \tag{2.1.5}$$

Note that in terms of a section,

$$\hat{\Delta} = \{e_\alpha,\ \kappa e_\alpha \mid \alpha \in \Delta\}. \tag{2.1.6}$$

Form the vector space $\underline{\mathfrak{h}} = L \otimes_{\mathbf{Z}} \mathbf{C}$, and extend $\langle \cdot, \cdot \rangle$ to $\underline{\mathfrak{h}} \times \underline{\mathfrak{h}}$ in the natural way. Set $\underline{\mathfrak{g}} = \underline{\mathfrak{h}} \oplus \sum_{a \in \hat{\Delta}} \mathbf{C} x_a$. (In terms of a section, $\underline{\mathfrak{g}} = \underline{\mathfrak{h}} \oplus \sum_{\alpha \in \Delta} \mathbf{C} x_\alpha$, where $x_\alpha = x_{e_\alpha}$.) The vector x_a is to be a nonzero element subject only to the relation $x_{\kappa a} = -x_a$. Define an alternating bilinear map $[\cdot, \cdot]$ on $\underline{\mathfrak{g}} \times \underline{\mathfrak{g}}$ as follows:

$$\begin{aligned}
[\underline{\mathfrak{h}}, \underline{\mathfrak{h}}] &= 0 \\
[h, \bar{a}] &= \langle h, \bar{a} \rangle x_a \\
[x_a, x_b] &= \begin{cases} \bar{a} & \text{if } ab = 1 \\ x_{ab} & \text{if } ab \in \hat{\Delta} \\ 0 & \text{if } ab \notin \hat{\Delta} \cup \{1, \kappa\}, \end{cases}
\end{aligned} \tag{2.1.7}$$

where $h \in \underline{\mathfrak{h}}$ and $a,\ b \in \hat{\Delta}$. (In terms of a section, we have

$$\begin{aligned}
[\underline{\mathfrak{h}}, \underline{\mathfrak{h}}] &= 0 \\
[h, x_\alpha] &= \langle h, \alpha \rangle x_\alpha \\
[x_\alpha, x_\beta] &= \begin{cases} \epsilon(\alpha,\beta)\alpha & \text{if } \alpha+\beta = 0,\ (\langle \alpha, \beta \rangle = -2), \\ \epsilon(\alpha,\beta) x_{\alpha+\beta} & \text{if } \alpha+\beta \in \Delta,\ (\langle \alpha, \beta \rangle = -1), \\ 0 & \text{if } \alpha+\beta \notin \Delta \cup 0, \end{cases}
\end{aligned} \tag{2.1.8}$$

where $h \in \underline{h}$, and α, $\beta \in \Delta$, and $\epsilon(\alpha,\beta) = (-1)^{\epsilon_0(\alpha,\beta)}$.)

Next extend $\langle\cdot,\cdot\rangle$ to a bilinear symmetric form on $\underline{g} \times \underline{g}$ by defining

$$\begin{aligned}
\langle\cdot,\cdot\rangle|_{\underline{h}\times\underline{h}} &= \langle\cdot,\cdot\rangle \\
\langle\underline{h}, x_a\rangle &= 0 \\
\langle x_a, x_b\rangle &= \begin{cases} 1 & \text{if } ab = 1, \\ 0 & \text{if } ab \notin \{1,\kappa\} \end{cases}
\end{aligned} \tag{2.1.9}$$

where a, $b \in \hat{\Delta}$. (In terms of a section e, we have $\langle x_\alpha, x_\beta\rangle = \epsilon(\alpha,\beta)\delta_{\alpha+\beta,0}$ for α, $\beta \in \Delta$.) We next state the following well-known result.

Theorem 2.1.1 ([F-K],[Seg]) *$(\underline{g}, [\cdot,\cdot])$ is a Lie algebra and the nonsingular form $\langle\cdot,\cdot\rangle$ is symmetric and $\underline{g}$-invariant. Furthermore, depending on the lattice type, $\underline{g}$ is isomorphic to A_l, D_l or E_l.*

The affine Kac-Moody algebra $\tilde{\underline{g}}$ associated with $(\underline{g}, \langle\cdot,\cdot\rangle)$ is the Lie algebra, which as a vector space is given by

$$\tilde{\underline{g}} = \underline{g} \otimes \mathbf{C}[t,t^{-1}] \oplus \mathbf{C}c \oplus \mathbf{C}d, \tag{2.1.10}$$

with commutation relations given by

$$\begin{aligned}
[x \otimes t^m, y \otimes t^n] &= [x,y] \otimes + \langle x,y\rangle m\delta_{m+n,\,0}c \\
[c, \tilde{\underline{g}}] &= [\tilde{\underline{g}}, c] = 0 \\
[d, x \otimes t^m] &= -[x \otimes t^m, d] = mx \otimes t^m
\end{aligned} \tag{2.1.11}$$

for x, $y \in \underline{g}$ and m, $n \in \mathbf{Z}$. The element c is called the central element and the element d is called the degree operator. We next extend the form $\langle\cdot,\cdot\rangle$ to a symmetric $\tilde{\underline{g}}$-invariant bilinear form on $\tilde{\underline{g}} \times \tilde{\underline{g}}$ by setting

$$\begin{aligned}
\langle x \otimes t^m, y \otimes t^n\rangle &= \langle x,y\rangle\delta_{m+n,\,0} \\
\langle c, x \otimes t^m\rangle &= \langle x \otimes t^m, c\rangle = 0 \\
\langle d, x \otimes t^m\rangle &= \langle x \otimes t^m, d\rangle = 0 \\
\langle c, d\rangle &= \langle d, c\rangle = 1 \\
\langle c, c\rangle &= \langle d, d\rangle = 0.
\end{aligned} \tag{2.1.12}$$

Note that the degree operator d gives a natural $\mathbf{Z}$-grading of $\tilde{\underline{g}}$: $\tilde{\underline{g}} = \coprod_{n\in\mathbf{Z}} \tilde{\underline{g}}_{[n]}$, where $\tilde{\underline{g}}_{[n]}$ is the ad d eigenspace with eigenvalue n. It is well-known that $\tilde{\underline{g}}$ is an affine Lie algebra of type $A_l^{(1)}$, $D_l^{(1)}$ or $E_l^{(1)}$ (cf. [M], [F-L-M]).

The Cartan subalgebra of $\tilde{\underline{g}}$ is given by

$$\underline{h}^e = \underline{h} \oplus \mathbf{C}c \oplus \mathbf{C}d. \tag{2.1.13}$$

The nonsingularity of $\langle\cdot,\cdot\rangle$ on $\Delta \times \Delta$ and (2.1.12) show that the form $\langle\cdot,\cdot\rangle$ is nondegenerate on the Cartan subalgebra $\underline{h}^e$. Thus we may identify $\underline{h}^e$ with its dual.

We say that $\alpha\,(\neq 0) \in \underline{h}^e$ is a root of $\tilde{\underline{g}}$ if the space

$$\tilde{\underline{g}}^\alpha = \{x \in \tilde{\underline{g}} \mid [h,x] = \langle h,\alpha\rangle x \text{ for all } \ h \in \underline{h}^e\}$$

contains at least one non-zero element of $\tilde{\underline{g}}$. By (2.1.11), (2.1.12) and (2.1.13) we see that the set of roots of $\tilde{\underline{g}}$ is given by

$$\begin{aligned} \Delta(\tilde{\underline{g}}) &= \{\alpha + mc \mid \alpha \in \Delta,\ m \in \mathbf{Z}\} \cup \{nc \mid n \in \mathbf{Z}\setminus\{0\}\} \\ &= \Delta_R(\tilde{\underline{g}}) \cup \Delta_I(\tilde{\underline{g}}). \end{aligned} \tag{2.1.14}$$

$\Delta_R(\tilde{\underline{g}})$ is called the *set of real roots of* $\tilde{\underline{g}}$ and $\Delta_I(\tilde{\underline{g}})$ is the *set of imaginary roots* of $\tilde{\underline{g}}$.

For $\alpha \in \Delta(\tilde{\underline{g}})$ with $\langle\alpha,\alpha\rangle \neq 0$, define $\alpha^\vee = \frac{2\alpha}{<\alpha,\alpha>}$. We next recall from [G] the definition of a Chevalley basis for the affine Lie algebra $\tilde{\underline{g}}$. Choose a section $e : L \to \hat{L}$. A *Chevalley basis* for the affine Lie algebra $\tilde{\underline{g}}$ is a basis of $\tilde{\underline{g}}$ of the form

$$\begin{aligned} &\{x_{\alpha+mc}\,(\in \tilde{\underline{g}}^{\alpha+mc}) \mid \alpha \in \Delta,\ n \in \mathbf{Z}\} \cup \{\alpha^\vee \otimes t^m \mid \alpha \in \Pi,\ m \in \mathbf{Z}\} \\ &\cup \{(\alpha_0)^\vee \mid \alpha_0 \text{ is the lowest root of } \tilde{\underline{g}}\} \cup \{d\} \end{aligned}$$

such that

1. $[x_{\alpha+mc}, x_{-\alpha-mc}] = \epsilon(\alpha,-\alpha)(\alpha+mc)^\vee$,

2. the linear map $\theta : \tilde{\underline{g}} \to \tilde{\underline{g}}$ defined by

$$x_{\alpha+mc} \ \mapsto\ -\epsilon(\alpha,-\alpha)x_{-\alpha-mc}$$

$$\begin{aligned} \alpha \otimes t^m &\mapsto -\alpha \otimes t^m \\ c &\mapsto -c \\ d &\mapsto -d, \end{aligned}$$

where $\alpha \in \Delta$ and $m \in \mathbf{Z}$, is a Lie algebra automorphism of $\tilde{\underline{g}}$.

The linear map θ is called a *Chevalley involution* of $\tilde{\underline{g}}$.

Using (2.1.8), (2.1.11) and (2.1.12) we have (cf. [G], [M]):

Theorem 2.1.2 *There exists a section* $e : L \to \hat{L}$ *such that* $e(0) = 1$ *and* $\epsilon_0(\alpha, \alpha) \equiv \frac{\langle \alpha, \alpha \rangle}{2} \bmod 2\mathbf{Z}$ *for all* $\alpha \in L$. *Choose such a section* e. *Then the set*

$$\begin{aligned} S &= \{x_\alpha \otimes t^m \mid \alpha \in \Delta,\, m \in \mathbf{Z}\} \cup \{\alpha_i \otimes t^m \mid 1 \le i \le l,\, m \in \mathbf{Z}\} \\ &\quad \cup \{\alpha_0 + c\} \cup \{d\} \end{aligned}$$

is a Chevalley basis of $\tilde{\underline{g}}$, *and* $\theta : \tilde{\underline{g}} \to \tilde{\underline{g}}$ *is an isometry of* $\tilde{\underline{g}}$ *with respect to the* $\tilde{\underline{g}}$-*invariant form* $\langle \cdot, \cdot \rangle$.

2.2 $B_n^{(1)}$ $(n \ge 2)$, $C_n^{(1)}$ $(n \ge 2)$, $F_4^{(1)}$

Let ν be a graph automorphism of the Dynkin diagram associated to an indecomposable root system of type A_{2n-1}, D_{n+1} or E_6 such that $\nu^2 = 1$, and ν is not the identity map. Extend ν linearly to Δ, L and $\underline{h}$. Since we are concerned only with the simply-laced root systems mentioned above, we are assured that

$$(-1)^{\langle \alpha, \nu\alpha \rangle} = 1 \tag{2.2.1}$$

for all $\alpha \in \Delta$.

Remark 2.2.1 Since ν is a linear extension of a graph automorphism, we see that the roots α not fixed by ν have the property $\langle \alpha, \nu\alpha \rangle = -1$ or 0. By (2.2.1) $\langle \alpha, \nu\alpha \rangle \in 2\mathbf{Z}$, and so we must have $\langle \alpha, \nu\alpha \rangle = 0$ whenver $\alpha \in \Delta$ and $\nu\alpha \neq \alpha$.

Because ν and its extensions are isometries, the automorphism ν (on L) lifts to an automorphism $\hat{\nu}$ on $\hat{L}$ such that $\overline{\hat{\nu}} = \nu$ (cf. Prop. 5.4.1, [F-L-M]). In fact we can, and do, select a lifting $\hat{\nu}$ of ν so that $\hat{\nu}a = a$ whenever $\nu\bar{a} = \bar{a}$ for $a \in \hat{L}$ (cf. §5, [L1]). In particular, this lifting of ν is such that $\hat{\nu}^2 = 1$ on $\hat{L}$.

Next we extend $\hat{\nu}$ to $\underline{g}$ by setting

$$\begin{aligned} \hat{\nu}|_{\underline{h}} &= \nu \\ \hat{\nu}x_a &= x_{\hat{\nu}a} \end{aligned} \tag{2.2.2}$$

for $a \in \hat{\Delta}$.

For a subset S of $\underline{g}$ and $n \in \mathbf{Z}$ set

$$S_{[n]} = \{x \in \underline{g} \mid \hat{\nu}x = (-1)^n x\}. \tag{2.2.3}$$

For $x \in \underline{g}$ and $n \in \mathbf{Z}$ we define

$$\begin{aligned} x_{[n]} &= \tfrac{1}{2}(\alpha + (-1)^n \hat{\nu}\alpha) \\ x^+_{[n]} &= \begin{cases} x_{[n]} & \text{if } x = (-1)^n \hat{\nu}x \\ 2x_{[n]} & \text{otherwise} \end{cases} \end{aligned} \tag{2.2.4}$$

We notice that the set of fixed points, $\Delta_{[0]}$, of the root system Δ of type A_{2n-1}, D_{n+1} or E_6, respectively, is a root system of type C_n, B_n or F_4, respectively (cf. [M]). Also the fixed point subalgebra $\underline{g}_{[0]}$ is a Lie algebra of type C_n, B_n or E_n. Form the Lie algebra $\widetilde{\underline{g}_{[0]}}$.

We may also extend $\hat{\nu}$ to a linear map of $\tilde{\underline{g}}$: for $x \in \underline{g}$, $n \in \mathbf{Z}$ and r, $s \in \mathbf{C}$ define

$$\hat{\nu}(x \otimes t^n + rc + sd) = \hat{\nu}x \otimes t^n + rc + sd. \tag{2.2.5}$$

It is not hard to see that $\hat{\nu}$ is an involutive automorphism of $\tilde{\underline{g}}$ and an isometry of $\tilde{\underline{g}}$ with respect to $\langle \cdot, \cdot \rangle$. Then we observe that

$$\tilde{\underline{g}}_{[0]} = \underline{g}_{[0]} \otimes \mathbf{C}[t, t^{-1}] \oplus \mathbf{C}c \oplus \mathbf{C}d;$$

so we have $\widetilde{\underline{\mathbf{g}}_{[0]}} = \tilde{\underline{\mathbf{g}}}_{[0]}$ (cf. [M]). Thus using (2.1.11), (2.2.2) and (2.2.4) we obtain the following theorem (cf. [F-L-M], [M]):

Theorem 2.2.2 *The algebra $\tilde{\underline{\mathbf{g}}}_{[0]}$ is isomorphic to either $C_n^{(1)}$, $B_n^{(1)}$ or $F_4^{(1)}$. Furthermore, $\tilde{\underline{\mathbf{g}}}_{[0]}$ has commutation relations given by:*

for $\alpha,\ \beta \in \Delta,\ b \in \hat{\Delta}$ and $m,\ n \in \mathbf{Z}$,

$$[\alpha_{[0]}^{+} \otimes t^m, \beta_{[0]}^{+} \otimes t^n] = m\langle \alpha_{[0]}^{+}, \beta_{[0]}^{+} \rangle \delta_{m+n,\,0}c; \tag{2.2.6}$$

$$[\alpha_{[0]}^{+} \otimes t^m, x_{b,\,[0]}^{+} \otimes t^n] = \langle \alpha_{[0]}^{+}, \overline{b}_{[0]} \rangle x_{b,\,[0]}^{+} \otimes t^{m+n}; \tag{2.2.7}$$

for $a,\ b \in \hat{\Delta}$ with $\langle \overline{a}_{[0]}, \overline{a}_{[0]} \rangle \leq \langle \overline{b}_{[0]}, \overline{b}_{[0]} \rangle$ and $\langle \overline{a}, \overline{b} \rangle \leq \langle \nu\overline{a}, \overline{b} \rangle$, and $m,\ n \in \mathbf{Z}$,

$$[x_{a,\,[0]}^{+} \otimes t^m, x_{b,\,[0]}^{+} \otimes t^n]$$

$$= \begin{cases} \overline{a}_{[0]}^{+} \otimes t^{m+n} + m\langle x_{a,\,[0]}^{+}, x_{a^{-1},\,[0]}^{+} \rangle \delta_{m+n,\,0}c & \text{if } b = a^{-1} \\ x_{ab,\,[0]}^{+} \otimes t^{m+n} & \text{if } \langle \overline{a}, \overline{b} \rangle = 1 \text{ and either} \\ & \overline{ab} \in \Delta^1, \text{ or } \overline{a},\ \overline{b},\ \overline{ab} \in \Delta^0, \\ 2x_{ab,\,[0]}^{+} \otimes t^{m+n} & \text{if } \langle \overline{a}, \overline{b} \rangle = -1 \text{ with} \\ & \overline{ab} \in \Delta^0 \text{ and } \overline{a},\ \overline{b} \in \Delta^1 \\ 0 & \text{if } \langle \overline{a}, \overline{b} \rangle \geq 0; \end{cases} \tag{2.2.8}$$

for $x \in \tilde{\underline{\mathbf{g}}}$ and $m \in \mathbf{Z}$,

$$[d, x_{[0]}^{+} \otimes t^m] = m x_{[0]}^{+} \otimes t^m. \tag{2.2.9}$$

Remark 2.2.3 The above bracket relations may also be written in terms of a chosen section $e : L \to \hat{L}$. However, the choice of section must be restricted to those maps with associated 2 cocycles ϵ_0 such that

$$\epsilon_0(\alpha, \beta) = \epsilon_0(\nu\alpha, \nu\beta) \tag{2.2.10}$$

for all $\alpha,\ \beta \in \Delta$. It is possible to choose such a section (cf. (3.2.3) of [M], (5.2.14) of [F-L-M]).

From (2.2.6), (2.2.7) and (2.2.9) of Theorem 2.2.2 we see that the Cartan subalgebra of $\tilde{\underline{g}}_{[0]}$ is given by

$$\underline{h}^e_{[0]} = \underline{h}_{[0]} \oplus \mathbf{C}c \oplus \mathbf{C}d, \tag{2.2.11}$$

and the set of roots for $\tilde{\underline{g}}_{[0]}$ is given by

$$\begin{aligned} \Delta(\tilde{\underline{g}}_{[0]}) &= \{\alpha_{[0]} + mc \mid \alpha \in \Delta,\ m \in \mathbf{Z}\} \cup \{nc \mid n \in \mathbf{Z} \setminus \{0\}\} \\ &= \Delta_R(\tilde{\underline{g}}_{[0]}) \cup \Delta_I(\tilde{\underline{g}}_{[0]}). \end{aligned} \tag{2.2.12}$$

Recall that ν is a graph automorphism of the Dynkin diagram associated to Δ. Hence we may also view ν as an element of the symmetric group on l elements. Define $I \subset \{1, 2, \ldots, l\}$ to be a set of representatives of the orbits of the automorphism ν on $\{1, 2, \ldots, l\}$. The set

$$\Pi_{[0]} = \{\alpha_{i,\,[0]} \mid i \in I\} \tag{2.2.13}$$

is linearly independent and spans $\Delta_{[0]}$; hence $\Pi_{[0]}$ is also a basis of $\Delta_{[0]}$.

Choose a section $e : L \to \hat{L}$ such that the associated 2 cocycle ϵ_0 safisfies (2.2.10). We take the definition of a Chevalley basis of $\tilde{\underline{g}}_{[0]}$ [G] to be a basis of $\tilde{\underline{g}}_{[0]}$ of the form

$$\{x_{\alpha_{[0]}+mc} \,(\in \tilde{\underline{g}}_{[0]}^{\alpha_{[0]}+mc}) \mid \alpha_{[0]} \in \Delta_{[0]},\ m \in \mathbf{Z}\} \cup \{\alpha^{\vee}_{[0]} \otimes t^m \mid \alpha_{[0]} \in \Pi_{[0]},\ m \in \mathbf{Z}\}$$
$$\cup \{(\alpha_{0,[0]} + c)^{\vee}\} \cup \{d\}$$

such that

1. $[x_{\alpha_{[0]}+mc}, x_{-\alpha_{[0]}-mc}] = \epsilon(\alpha, -\alpha)(\alpha_{[0]} + mc)^{\vee}$,

2. the linear map $\theta|_{\tilde{\underline{g}}_{[0]}}$ is a Lie algebra automorphism of $\tilde{\underline{g}}_{[0]}$.

Note that using Remark 2.2.1 it is easy to see that we have

$$(\alpha_{[0]} + mc)^{\vee} = \begin{cases} \alpha^{+}_{[0]} + mc & \text{if } \alpha \in \Delta^0, \\ \alpha^{+}_{[0]} + 2mc & \text{if } \alpha \in \Delta^1. \end{cases} \tag{2.2.14}$$

Thus Theorem 2.2.2 and (2.2.14) imply that

$$[x^{+}_{\alpha_{[0]}} \otimes t^m, x^{+}_{-\alpha_{[0]}} \otimes t^{-m}] = \epsilon(\alpha, -\alpha)(\alpha_{[0]} + mc)^{\vee}$$

for an appropriately chosen section (cf. (2.2.10)). We now have the analogue to Theorem 2.1.2 for the algebras $B_n^{(1)}$, $C_n^{(1)}$ and $F_4^{(1)}$ (cf. [G], [M]). Let $\check{\Delta}$ be a set of representatives of the orbits of Δ under ν.

Theorem 2.2.4 *Let $e : L \to \hat{L}$ be a section so that the corresponding 2-cocycle ϵ_0 satisfies both (2.2.10) for $\alpha \in \Delta$ and $\epsilon_0 \equiv \frac{\langle \alpha, \alpha \rangle}{2} \bmod 2\mathbf{Z}$ for all $\alpha \in L$ (cf. [F-L-M]). Then the set*

$$\begin{aligned} S^+_{[0]} &= \{x^+_{[0]} \otimes t^n \mid \alpha \in \check{\Delta},\, n \in \mathbf{Z}\} \cup \{\alpha^+_{i,[0]} \otimes t^n \mid i \in I,\, n \in \mathbf{Z}\} \\ &\quad \cup \{\alpha^+_{0,[0]} + c\} \cup \{d\} \end{aligned}$$

is a Chevalley basis of $\tilde{\underline{\mathfrak{g}}}_{[0]}$, and the involution θ is an isometry with respect to the $\tilde{\underline{\mathfrak{g}}}_{[0]}$-invariant form $\langle \cdot, \cdot \rangle$.

2.3 $A_{2n-1}^{(2)}$ $(n \geq 2)$, $D_{n+1}^{(2)}$ $(n \geq 2)$, $E_6^{(2)}$

To construct the affine Lie algebras of type $A_{2n-1}^{(2)}$, $D_{n+1}^{(2)}$ $(n \geq 2)$, or $E_6^{(2)}$, we consider another fixed point subalgebra of $\tilde{\underline{\mathfrak{g}}}$. As in [M], define an automorphism τ of $\tilde{\underline{\mathfrak{g}}}$ by

$$\tau(x \otimes t^n + rc + sd) = (-1)^n \hat{\nu} x \otimes t^n + rc + sd \tag{2.3.1}$$

where $x \in \underline{\mathfrak{g}}$, $n \in \mathbf{Z}$, $\hat{\nu}$ is given by (2.2.2), and $r, s \in \mathbf{C}$. The automorphism τ is involutive and an isometry with respect to the form $\langle \cdot, \cdot \rangle$. The fixed subalgebra $\tilde{\underline{\mathfrak{g}}}^{(\tau)}$ of $\tilde{\underline{\mathfrak{g}}}$ is given by

$$\tilde{\underline{\mathfrak{g}}}^{(\tau)} = \coprod_{n \in \mathbf{Z}} \tilde{\underline{\mathfrak{g}}}_{[n]} \oplus \mathbf{C}c \oplus \mathbf{C}d = \coprod_{n \in \mathbf{Z}} \left(\underline{\mathfrak{g}}_{[n]} \otimes t^n \right) \oplus \mathbf{C}c \oplus \mathbf{C}d$$

and is isomorphic to one of the above types of affine Lie algebras (cf. [M]). Moreover, the form $\langle \cdot, \cdot \rangle$ on $\tilde{\underline{\mathfrak{g}}}$ restricts to a nondegenerate $\tilde{\underline{\mathfrak{g}}}$-invariant form on the fixed point subalgebra of $\tilde{\underline{\mathfrak{g}}}$. We record these facts along with the commutation relations for these affine algebras in the following (cf. [M]):

Theorem 2.3.1 *The Lie algebra $\underline{\tilde{g}}^{(\tau)}$ is isomorphic to either $A_{2n-1}^{(2)}$, $D_{n+1}^{(2)}$ or $E_6^{(2)}$. Furthermore the algebra $\underline{\tilde{g}}^{(\tau)}$ has commutation relations given by:*

for $\alpha,\ \beta \in \Delta,\ b \in \hat{\Delta}$, and $m,\ n \in \mathbf{Z}$

$$[\alpha^+_{[m]}\otimes t^m, \beta^+_{[n]}\otimes t^n] = \langle \alpha^+_{[m]}, \beta^+_{[n]}\rangle m\delta_{m+n,\,0}c; \tag{2.3.2}$$

$$[\alpha^+_{[m]}\otimes t^m, x^+_{b,\,[n]}\otimes t^n] = \langle \alpha^+_{[m]}, \overline{b}_{[n]}\rangle x^+_{b,\,[m+n]}\otimes t^{m+n}; \tag{2.3.3}$$

for $a,\ b \in \hat{\Delta}$ such that $\langle \overline{a}, \overline{b}\rangle \le \langle \nu\overline{a}, \overline{b}\rangle$ and $m,\ n \in \mathbf{Z}$,

$$[x^+_{a,\,[m]}\otimes t^m, x^+_{b,\,[n]}\otimes t^n]$$

$$= \begin{cases} \overline{a}^+_{[m+n]}\otimes t^{m+n} + \langle x^+_{a,\,[m]}, x^+_{a^{-1},\,[-m]}\rangle m\delta_{m+n,\,0}c & \text{if } b = a^{-1} \\ x^+_{ab,\,[m+n]}\otimes t^{m+n} & \text{if } \langle \overline{a}, \overline{b}\rangle = -1 \text{ and} \\ & \text{either } \overline{ab} \in \Delta^1 \\ & \text{or } \overline{a},\ \overline{b},\ \overline{ab} \in \Delta^0, \\ 2x^+_{ab,\,[m+n]}\otimes t^{m+n} & \text{if } \langle \overline{a}, \overline{b}\rangle = -1, \text{ with} \\ & \overline{ab} \in \Delta^0 \text{ and } \overline{a},\ \overline{b} \in \Delta^1, \\ 0 & \text{if } \langle \overline{a}, \overline{b}\rangle \ge 0; \end{cases} \tag{2.3.4}$$

for $x \in \underline{g}$ and $m \in \mathbf{Z}$,

$$[d, x^+_{[m]}\otimes t^m] = m x^+_{[m]}\otimes t^m. \tag{2.3.5}$$

Remark 2.3.2 We also obtain analogous bracket relations in terms of a carefully chosen section e: e must be selected so that its associated 2-cocycle ϵ satisfies (2.2.10) (cf. Remark 2.2.3).

Using (2.3.2), (2.3.3) and (2.3.5) of Theorem 2.3.1 we find that the Cartan subalgebra of $\underline{\tilde{g}}^{(\tau)}$ is once again given by (2.2.11), while the set of roots of $\underline{\tilde{g}}^{(\tau)}$ is given by

$$\begin{aligned} \Delta(\underline{\tilde{g}}^{(\tau)}) &= \{\alpha_{[0]} + 2mc \mid \alpha \in \Delta,\ m \in \mathbf{Z}\} \cup \{\beta_{[0]} + (2m+1)c \mid \beta \in \Delta^1,\ m \in \mathbf{Z}\} \\ &\quad \cup \{nc \mid n \in \mathbf{Z} \setminus \{0\}\} \\ &= \Delta_R(\underline{\tilde{g}}^{(\tau)}) \cup \Delta_I(\underline{\tilde{g}}^{(\tau)}). \end{aligned} \tag{2.3.6}$$

Choose a section $e : L \to \hat{L}$ such that the associated 2 cocycle satisfies (2.2.10). A *Chevalley basis* of $\tilde{\mathbf{g}}^{(\tau)}$ [M] is a basis of $\tilde{\mathbf{g}}^{(\tau)}$ of the form

$$\begin{aligned}&\{x_{\alpha_{[0]}+2mc}\left(\in (\tilde{\mathbf{g}}^{(\tau)})^{\alpha_{[0]}+2mc}\right) \mid \alpha \in \Delta,\ m \in \mathbf{Z}\}\\ &\quad \cup \{x_{\beta_{[0]}+(2m+1)c}\left(\in (\tilde{\mathbf{g}}^{(\tau)})^{\beta_{[0]}+(2m+1)c}\right) \mid \beta \in \Delta^1,\ m \in \mathbf{Z}\}\\ &\quad \cup \{\alpha_{[0]}^{\vee}\otimes t^{2m} \mid \alpha_{[0]} \in \Pi_{[0]},\ m \in \mathbf{Z}\} \cup \{\beta_{[1]}^{\vee}\otimes t^{2m+1} \mid \beta \in \Pi_{[0]} \cap \Delta^1,\ m \in \mathbf{Z}\}\\ &\quad \cup \{\left(\gamma_{[0]} + c\right)^{\vee}\} \cup \{d\},\end{aligned}$$

(where $\gamma \in \Delta^1$ such that $\gamma_{[0]}$ is the lowest weight of the adjoint $\mathbf{g}_{[0]}$-module $\mathbf{g}_{[1]}$) such that

1. $[x_{\alpha_{[0]}+mc}, x_{-\alpha_{[0]}-mc}] = \epsilon(\alpha, -\alpha)(\alpha_{[0]} + mc)^{\vee}$ for $\alpha_{[0]} + mc \in \Delta_R(\tilde{\mathbf{g}}^{(\tau)})$,

2. the linear map $\theta|_{\tilde{\mathbf{g}}^{(\tau)}}$ is a Lie algebra automorphism of $\tilde{\mathbf{g}}^{(\tau)}$.

It follows from (2.2.14) and Theorem 2.3.1 that

$$[x^{+}_{\alpha,[m]} \otimes t^{m}, x^{+}_{-\alpha,[-m]} \otimes t^{-m}] = \epsilon(\alpha, -\alpha)(\alpha_{[0]} + mc)^{\vee}$$

for $\alpha \in \check{\Delta}$ and $m \in \mathbf{Z}$. Thus we have the following:

Theorem 2.3.3 ([M]) *Choose a section $e : L \to \hat{L}$ and a corresponding cocycle ϵ_0 as in Theorem 2.2.4. Then*

$$\begin{aligned}S^{(\tau)} \;=\; &\{x^{+}_{\alpha,[2m]}\otimes t^{2m} \mid \alpha \in \check{\Delta}\} \cup \{x^{+}_{\beta,[2m+1]}\otimes t^{2m+1} \mid \beta \in \check{\Delta} \cap \Delta^1,\ m \in \mathbf{Z}\}\\ &\cup \{\alpha_{[0]}^{+}\otimes t^{2m} \mid \alpha_{[0]} \in \Delta_{[0]},\ m \in \mathbf{Z}\} \cup \{\beta_{[1]}^{+}\otimes t^{2m+1} \mid \beta \in \Pi_{[0]} \cap \Delta^1,\ m \in \mathbf{Z}\}\\ &\cup \{\gamma_{[0]}^{+} + 2c\} \cup \{d\}\end{aligned}$$

is a Chevalley basis of $\tilde{\mathbf{g}}^{(\tau)}$ and θ is an isometry with respect to the $\tilde{\mathbf{g}}^{(\tau)}$-invariant form $\langle\cdot,\cdot\rangle$.

2.4 $G_2^{(1)}$

In this section we restrict Δ to be the root system for the Lie algebra D_4. Construct L, $(\hat{L}, -)$, and c_0 as in §2.1.1. Assume Π is given by $\{\alpha_1, \alpha_2, \alpha_3, \alpha_4\}$ where $\langle\alpha_i, \alpha_2\rangle = -1$

for $i = 1, 3, 4$, and $\langle \alpha_i, \alpha_j \rangle = 0$ whenever $i \neq j$ and $i, j = 1, 3, 4$. We next define ν to be the graph automorphism of order three given by

$$\nu\alpha_1 = \alpha_3,\ \nu\alpha_2 = \alpha_2,\ \nu\alpha_3 = \alpha_4 \text{ and } \nu\alpha_4 = \alpha_1. \tag{2.4.1}$$

Extend ν linearly to Δ, L, and $\underline{h}$. Since ν and its extensions are isometries, we may extend ν to an automorphism $\hat{\nu}$ of $\hat{L}$ such that $\overline{\hat{\nu}} = \nu$ on L (Prop. 5.4.1, [F-L-M]). Furthermore, we can choose an extension $\hat{\nu}$ on $\hat{L}$ so that, in addition to the above property, $\hat{\nu}^3 = 1$ on $\hat{L}$ and $\hat{\nu}a = a$ whenever $\nu\overline{a} = \overline{a}$ (cf. §5, [L1]). Let $\hat{\nu}$ be such an extension of ν. Next extend $\hat{\nu}$ to and automorphism of $\underline{g}$ by setting $\hat{\nu}x_a = x_{\hat{\nu}a}$.

Let ω be a primitive third root of unity. For a subset S of $\underline{g}$, set

$$S_{[n]} = \{x \in \underline{g} \mid \hat{\nu}x = \omega^n x\}. \tag{2.4.2}$$

For $x \in \underline{g}$ and $n \in \mathbf{Z}$, define $x_{[n]}$ and $x^+_{[n]}$ as follows:

$$\begin{aligned} x_{[n]} &= \tfrac{1}{3}(x + \omega^{-n}\hat{\nu}x + \omega^n\hat{\nu}^2 x) \\ x^+_{[n]} &= \begin{cases} x_{[n]} & \text{if } \hat{\nu}x = \omega^n x \\ 0 & \text{if } x \in \underline{g}_{[n-1]} \oplus \underline{g}_{[n+1]} \\ 3x_{[n]} & \text{otherwise.} \end{cases} \end{aligned} \tag{2.4.3}$$

Extend $\hat{\nu}$ to $\tilde{\underline{g}}$ by defining

$$\hat{\nu}(x \otimes t^n + rc + sd) = (\hat{\nu}x) \otimes t^n + rc + sd, \tag{2.4.4}$$

and give analogous definitions to (2.4.2) and (2.4.3) for $\tilde{\underline{g}}$. Note that the Lie algebras $\widetilde{\underline{g}_{[0]}}$ and $\tilde{\underline{g}}_{[0]}$ are the same.

For the root system D_4, we record a property which will be useful in later calculations.

Proposition 2.4.1 *If $\nu\alpha \neq \alpha$, then* $\langle \nu\alpha, \alpha \rangle = \langle \nu^2\alpha, \alpha \rangle = 0$.

Proof: First $\langle \nu\alpha, \alpha \rangle = \langle \nu^2\alpha, \alpha \rangle$ for all $\alpha \in \Delta$ since ν is an isometry with respect to $\langle \cdot, \cdot \rangle$. By definition ν is an extension of a graph automorphism, and so ν preserves the positive root system of Δ. Thus $\langle \nu^i\alpha, \alpha\rangle$ can only take the values 2, ± 1 or 0 for $i = 0, 1, 2$.

If $\langle \nu\alpha, \alpha\rangle = 2$, then $\nu\alpha = \nu^2\alpha = \alpha$. However, we assumed $\alpha \in \Delta^1$; so this is impossible. Next if $\langle \nu\alpha, \alpha\rangle = -1$, then we have $\nu\alpha + \alpha \in \Delta$. We see that this implies that $\langle \nu\alpha + \alpha, \nu^2\alpha\rangle = -2$, or equivalently, that $\nu\alpha + \alpha = -(\nu^2\alpha)$. Again we obtain a contradiction since ν preserves the positive root system of Δ. If $\langle \nu\alpha, \alpha\rangle = 1$, then we have $\nu\alpha - \alpha \in \Delta \setminus \Delta^0$. We next observe that $\langle \nu(\nu\alpha - \alpha), \nu\alpha - \alpha\rangle = -1$; but this was just shown to be impossible. Hence we must have $\langle \nu\alpha, \alpha\rangle = \langle \nu^2\alpha, \alpha\rangle = 0$ whenever $\alpha \notin \Delta^0$. □

Using case by case arguments which utilize Proposition 2.4.1, (2.1.7), (2.1.11) and (2.1.12), we are able to obtain the commutation relations for $\tilde{\mathfrak{g}}_{[0]}$ (cf. [M]).

Theorem 2.4.2 *The Lie algebra $\mathfrak{g}_{[0]}$ is isomorphic to $G_2^{(1)}$ and has the following commutation relations:*

for $\alpha,\ \beta \in \Delta,\ b \in \hat{\Delta}$ and $m,\ n \in \mathbf{Z}$

$$[\alpha^+_{[0]} \otimes t^m, \beta^+_{[0]} \otimes t^n] = \langle \alpha^+_{[0]}, \beta^+_{[0]}\rangle m\delta_{m+n,\,0}c, \tag{2.4.5}$$

$$[\alpha^+_{[0]} \otimes t^m, x^+_{b,\,[0]} \otimes t^n] = \langle \alpha^+_{[0]}, \overline{b}_{[0]}\rangle x_{b,\,[0]} \otimes t^{m+n}; \tag{2.4.6}$$

for $a,\ b \in \hat{\Delta}$ such that $\langle \overline{a}, \overline{b}\rangle \le \langle \nu\overline{a}, \overline{b}\rangle \le \langle \nu^2\overline{a}, \overline{b}\rangle$, and $m,\ n \in \mathbf{Z}$

$$[x^+_{a,\,[0]} \otimes t^m, x^+_{b,\,[0]} \otimes t^n]$$

$$= \begin{cases} \overline{a}^+_{[0]} \otimes t^{m+n} + \langle x^+_{a,\,[0]}, x^+_{a^{-1},\,[0]}\rangle m\delta_{m+n,\,0}c & \text{if } b = a^{-1} \\ x^+_{ab,\,[0]} \otimes t^{m+n} & \text{if } \langle \overline{a}, \overline{b}\rangle = -1 \text{ with } \\ & \quad \overline{a} \text{ or } \overline{b} \in \Delta^0, \\ 2x^+_{ab,\,[0]} \otimes t^{m+n} & \text{if } \langle \overline{a}, \overline{b}\rangle = -1, \text{ with } \\ & \quad \overline{a},\ \overline{b},\ \overline{ab} \in \Delta^1, \\ 3x^+_{ab,\,[0]} \otimes t^{m+n} & \text{if } \langle \overline{a}, \overline{b}\rangle = -1, \text{ with } \\ & \quad \overline{ab} \in \Delta^0 \text{ and } \overline{a},\ \overline{b} \in \Delta^1, \\ 0 & \text{if } \langle \overline{a}, \overline{b}\rangle \ge 0; \end{cases} \tag{2.4.7}$$

for $x \in \underline{\mathfrak{g}}$ *and* $m \in \mathbf{Z}$,

$$[d, x^+_{[0]} \otimes t^m] = m x^+_{[0]} \otimes t^m. \tag{2.4.8}$$

Remark 2.4.3 The preceding bracket relations have analogous versions in terms of a section e if e is chosen so that (2.2.10) is satisfied.

Theorem 2.4.2 shows that the Cartan subalgebra of $\tilde{\underline{\mathfrak{g}}}_{[0]}$ is again given by (2.2.11), and that the root system of $\tilde{\underline{\mathfrak{g}}}_{[0]}$ is given by (2.2.12).

Fix $j = 1, 3$ or 4. Define $\Pi_{[0]} = \{\alpha_{i,[0]} \mid i = 2, j\}$. Clearly $\Pi_{[0]}$ is a basis of $\Delta_{[0]}$. Next let $\check{\Delta}$ be the subset of Δ consisting of the roots $\pm\alpha_j$, $\pm\alpha_2$, $\pm(\alpha_2 + \alpha_j)$, $\pm(\alpha_2 + \nu\alpha_j + \nu^2\alpha_j)$, $\pm(\alpha_2 + \alpha_j + \nu\alpha_j + \nu^2\alpha_j)$, and $\pm(2\alpha_2 + \alpha_j + \nu\alpha_j + \nu^2\alpha_j)$. Thus we see that $\check{\Delta}$ is a set of representatives of the orbits of Δ under ν.

Next we state a characterization of two roots α, $\beta \in \check{\Delta}^1$ which will be useful in deriving commutation relations both in this section and in later sections.

Proposition 2.4.4 *(1) If* α, $\beta \in \check{\Delta}^1$, *then* $\alpha + \beta \notin \Delta^1$ *and* $\langle \alpha, \beta \rangle \neq 0$.
(2) If $\alpha, \beta \in \check{\Delta}^1$, *then* $\langle \alpha, \beta \rangle = 1$ *if and only if* $\langle \nu\alpha, \beta \rangle = -1$, *in which case* $\nu\alpha + \beta \in \check{\Delta}^1$.

Proof: These properties are checked by enumerating the pairs (α, β) with α, $\beta \in \check{\Delta}^1$, and $\langle \alpha, \beta \rangle = -1, 0, 1$ or $\langle \nu\alpha, \beta \rangle = -1$. □

We take the definition of a Chevalley basis of $\tilde{\underline{\mathfrak{g}}}_{[0]}$ to be the same definition found in §2.2 with the index set I replaced by the set $J = \{2, j\}$.

Proposition 2.4.1 implies that $\alpha^\vee_{[0]} = \alpha^+_{[0]}$ for $\alpha \in \Delta$. Thus

$$(\alpha_{[0]} + mc)^\vee = \begin{cases} \alpha^+_{[0]} + mc & \text{if } \alpha \in \Delta^0 \\ \alpha^+_{[0]} + 3mc & \text{if } \alpha \in \Delta^1. \end{cases} \tag{2.4.9}$$

Using Theorem 2.4.2 and (2.4.9) we have (cf. [G], [M]):

Theorem 2.4.5 *Choose a section* $e : L \to \hat{L}$ *and a corresponding cocycle* ϵ_0 *as in Theorem 2.2.4. Then*

$$\begin{aligned} S^+_{[0]} &= \{x^+_{\alpha,[0]} \otimes t^m \mid \alpha \in \check{\Delta},\ m \in \mathbf{Z}\} \cup \{\alpha^+_{i,[0]} \otimes t^m \mid i \in J,\ m \in \mathbf{Z}\} \\ &\quad \cup \{\alpha^+_{0,[0]} + c\} \cup \{d\} \end{aligned}$$

(where α_0 is the lowest root of $\underline{g}$) is a Chevalley basis of $\tilde{\underline{g}}_{[0]}$, and the involution θ of $\tilde{\underline{g}}_{[0]}$ is an isometry with respect to the $\tilde{\underline{g}}_{[0]}$-invariant form $\langle\cdot,\cdot\rangle$.

2.5 $D_4^{(3)}$

We retain the notation set up in the previous section.

To construct $D_4^{(3)}$, we define the following automorphism τ of $\tilde{\underline{g}}$:

$$\tau(x\otimes t^m + rc + sd) = (\omega)^{-m}(\hat{\nu}x)\otimes t^m + rc + sd \tag{2.5.1}$$

where $x \in \underline{g}$, r, $s \in \mathbf{C}$, $m \in \mathbf{Z}$ and $\hat{\nu}$ is the extension of ν to $\underline{g}$ (cf. §2.4). From (2.4.3) and (2.5.1) we see that the form $\langle\cdot,\cdot\rangle$ restricts to a nondegenerate $\tilde{\underline{g}}$-invariant form of the τ-fixed subalgebra $\tilde{\underline{g}}^{(\tau)}$, where $\tilde{\underline{g}}^{(\tau)}$ is given by

$$\tilde{\underline{g}}^{(\tau)} = \coprod_{n\in\mathbf{Z}} \tilde{\underline{g}}_{[n]} = \coprod_{n\in\mathbf{Z}} \underline{g}_{[n]}\otimes t^n \oplus \mathbf{C}c \oplus \mathbf{C}d.$$

Using (2.1.7), (2.1.11), (2.1.12), (2.4.3), Proposition 2.4.1 and Proposition 2.4.4, we have the following (cf. [M]):

Theorem 2.5.1 *The Lie algebra $\tilde{\underline{g}}^{(\tau)}$ is isomorphic to $D_4^{(3)}$ and has commutation relations given by: for α, $\beta \in \check{\Delta}$, $b \in \hat{\Delta}$ and $m, n \in \mathbf{Z}$*

$$[\alpha^+_{[m]}\otimes t^m, \beta^+_{[n]}\otimes t^n] = \langle\alpha^+_{[m]}, \beta^+_{[n]}\rangle m\delta_{m+n,\,0}c; \tag{2.5.2}$$

$$[\alpha^+_{[m]}\otimes t^m, x^+_{b,\,[n]}\otimes t^n] = \langle\alpha^+_{[m]}, \bar{b}_{[-m]}\rangle x^+_{b,\,[m+n]}\otimes t^{m+n}; \tag{2.5.3}$$

for a, $b \in \hat{\Delta}$, and m, $n \in \mathbf{Z}$ such that $x^+_{a,\,[m]}\otimes t^m \neq 0 \neq x^+_{b,\,[n]}\otimes t^n$

$$[x^+_{a,\,[m]}\otimes t^m, x^+_{b,\,[n]}\otimes t^n]$$

$$= \begin{cases} \overline{a}^+_{[m+n]}\otimes t^{m+n} + \langle x^+_{a,\,[m]}, x^+_{a^{-1},\,[-m]}\rangle m\delta_{m+n,\,0}c & \textit{if } b = a^{-1} \\ x^+_{ab,\,[m+n]}\otimes t^{m+n} & \textit{if } \langle \overline{a}, \overline{b}\rangle = -1 \textit{ with} \\ & \overline{a} \textit{ or } \overline{b} \in \check{\Delta}^0, \\ -x^+_{(\hat{\nu}^2 a)(\hat{\nu} b),\,[m+n]}\otimes t^{m+n} & \textit{if } \langle \nu^2\overline{a}, \nu\overline{b}\rangle = -1 \textit{ with} \\ & \overline{a},\, \overline{b},\, \nu^2\overline{a} + \nu\overline{b} \in \check{\Delta}^1 \\ & \textit{and } [m] \neq [n], \\ 2x^+_{(\hat{\nu}^2 a)(\hat{\nu} b),\,[m+n]}\otimes t^{m+n} & \textit{if } \langle \nu^2\overline{a}, \nu\overline{b}\rangle = -1 \textit{ with} \\ & \overline{a},\, \overline{b},\, \nu^2\overline{a} + \nu\overline{b} \in \check{\Delta}^1, \\ & \textit{and } [m] = [n], \\ 3x^+_{ab,\,[m+n]}\otimes t^{m+n} & \textit{if } \langle \overline{a}, \overline{b}\rangle = -1 \textit{ with} \\ & \overline{ab} \in \check{\Delta}^0 \textit{ and } \overline{a},\, \overline{b} \in \check{\Delta}^1, \\ 0 & \textit{if } \langle \overline{a}, \overline{b}\rangle \geq 0 \textit{ with } \overline{a} \\ & \textit{or } \overline{b} \in \check{\Delta}^0 \textit{ or if } \overline{a} = \overline{b}; \end{cases} \tag{2.5.4}$$

for $x \in \underline{\mathbf{g}}$ *and* $m \in \mathbf{Z}$

$$[d, x^+_{[m]}\otimes t^m] = mx^+_{[m]}\otimes t^m. \tag{2.5.5}$$

Remark 2.5.2 We may rewrite the above bracket relations in terms of a section e which satisfies (2.2.10).

We see from (2.5.2), (2.5.3) and (2.5.5) of Theorem 2.5.1 and the definition of $\langle\cdot,\cdot\rangle$ on $\tilde{\underline{\mathbf{g}}}^{(\tau)}$, that the Cartan subalgebra of $\tilde{\underline{\mathbf{g}}}^{(\tau)}$ is given by

$$\begin{aligned} \Delta(\tilde{\underline{\mathbf{g}}}^{(\tau)}) &= \{\alpha_{[0]} + 3mc \mid \alpha \in \check{\Delta},\ m \in \mathbf{Z}\} \cup \{\beta_{[0]} + (3m+1)c \mid \beta \in \check{\Delta}^1,\ m \in \mathbf{Z}\} \\ &\quad \cup \{nc \mid n \in \mathbf{Z} \setminus \{0\}\}. \end{aligned}$$

We next choose a section $e : L \to \hat{L}$ so that e satisfies (2.2.10). Then a *Chevalley basis* of $\tilde{\underline{\mathbf{g}}}^{(\tau)}$ [M] is a basis of $\tilde{\underline{\mathbf{g}}}^{(\tau)}$ of the form

$$\{\overline{x}^+_{\alpha_{[0]}+3mc} \left(\in (\tilde{\underline{\mathbf{g}}}^{(\tau)})^{\alpha_{[0]}+3mc}\right) \mid \alpha \in \Delta \text{ and } m \in \mathbf{Z}\}$$

$$\cup \{\overline{x}^{+}_{\beta_{[0]}+(3m\pm1)c} \left(\in (\tilde{\underline{g}}^{(\tau)})^{\beta_{[0]}+(3m\pm1)c}\right) \mid \beta \in \Delta^1 \text{ and } m \in \mathbf{Z}\}$$
$$\cup \{\alpha^{+}_{i,\,[0]} \otimes t^{3m} \mid m \in \mathbf{Z} \text{ and } i = 2,\ j\} \cup \{\alpha^{+}_{j,\,[3m\pm1]} \otimes t^{3m\pm1} \mid m \in \mathbf{Z}\}$$
$$\cup \{(\gamma^{+}_{[0]} + c)^{\vee}\} \cup \{d\}$$

where $\gamma \in \Delta^1$ is such that $\gamma_{[0]}$ is the lowest weight of the $\underline{g}_{[0]}$-modules $\underline{g}_{[\pm1]}$ where conditions 1 and 2 of the definition fould in §2.3 hold.

We next note that Theorem 2.5.1 and (2.4.9) imply

$$[\overline{x}^{+}_{\alpha,\,[m]} \otimes t^{m}, \overline{x}^{+}_{-\alpha,\,[-m]} \otimes t^{-m}] = \epsilon(\alpha, -\alpha)(\alpha_{[0]} + mc)^{\vee}$$

for an appropriate section e (cf. (2.2.10)). Thus we have (cf. [M]):

Theorem 2.5.3 *Choose a section e and corresponding 2-cocycle ϵ_0 as in Theorem 2.2.4. Then*

$$\mathbf{S}^{(\tau)} = \{\overline{x}^{+}_{\alpha,\,[0]} \otimes t^{3m} \mid \alpha \in \check{\Delta} \text{ and } m \in \mathbf{Z}\} \cup \{\overline{x}^{+}_{\beta,\,[\pm1]} \otimes t^{3m\pm1} \mid \beta \in \check{\Delta}^1 \text{ and } m \in \mathbf{Z}\}$$
$$\cup \{\alpha^{+}_{i,\,[0]} \otimes t^{3m} \mid i = j,\ 2 \text{ and } m \in \mathbf{Z}\} \cup \{\alpha^{+}_{j,\,[\pm1]} \otimes t^{3m\pm1} \mid m \in \mathbf{Z}\}$$
$$\cup \{\gamma^{+}_{[0]} + 3c\} \cup \{d\}$$

is a Chevalley basis of $\tilde{\underline{g}}^{(\tau)}$ and the involution θ of $\tilde{\underline{g}}^{(\tau)}$ is an isometry with respect to the $\tilde{\underline{g}}^{(\tau)}$-invariant form $\langle\cdot,\cdot\rangle$.

2.6 $A_{2n}^{(2)}$ $(n \geq 1)$

Let Δ be indecomposable of type A_{2n} with $n \geq 1$. Again form $(\hat{L}, -)$, $\underline{h}$, $\underline{g}$, and $\tilde{\underline{g}}$ as in §2.1. Define a graph automorphism ν of the associated Dynkin diagram so that the following conditions hold:

1. $(-1)^{\langle \nu\alpha, \alpha \rangle} = -1$ for at least one $\alpha \in \Delta$ and

2. $\nu^2 = 1$ and ν is not the identity map.

By indexing the simple roots $\Pi = \{\alpha_1, \ldots, \alpha_{2n}\}$ so that for $i \neq j$

$$\langle \alpha_i, \alpha_j \rangle = \begin{cases} -1 & \text{if } i = j \pm 1 \\ 0 & \text{otherwise,} \end{cases}$$

we see that the graph involution ν is given by $\nu\alpha_j = \alpha_{2n+1-j}$ for $j = 1, \ldots, 2n$. As before extend ν linearly to Δ and $\underline{h}$. Note that since ν and its extensions are isometries, we have $c_0(\alpha, \beta) = c_0(\nu\alpha, \nu\beta) = c_0(-\nu\alpha, -\nu\beta)$ for $\alpha,\ \beta \in \Pi$. Using Proposition 5.4.1 of [F-L-M] we can extend ν to an automorphism η of $\hat{L}$ so that $\overline{\eta} = -\nu$ on L.

We next define an antiautomorphism ν' of $\hat{L}$ given by

$$\nu'(a) = \eta(a^{-1}) \text{ for } a \in \hat{L}. \tag{2.6.1}$$

Thus we have $\overline{\nu'(a)} = \overline{\eta(a^{-1})} = -\nu(\overline{a^{-1}}) = -(\nu(-\overline{a})) = \nu(\overline{a})$. Also notice that since η fixes κ, $\nu'(\kappa) = \kappa$.

We next show that we can choose $\hat{\nu}$ to be an antiautomorphism so that $\overline{\hat{\nu}^2} = 1$ and $\overline{\hat{\nu}} = \nu$ on L. Let η be an automorphism of L as above. Then for any $a \in \hat{L}$ we observe that $\eta^2(a) = \kappa^{u(a)}a$ where $u : \hat{L} \longrightarrow \mathbf{Z}/2\mathbf{Z}$ is a homomorphism. Next we see that $\eta^2(\kappa a) = \kappa^{u(\kappa a)}(\kappa a)$. Since η fixes κ, we also must have $\eta^2(\kappa a) = \kappa\eta^2(a) = \kappa^{u(a)}(\kappa a)$. Thus $u(\kappa a) = u(a)$ for all $a \in \hat{L}$. Now define $\overline{u} : L \longrightarrow \mathbf{Z}/2\mathbf{Z}$ by $\overline{u}(\overline{a}) = u(a)$. Note that $\overline{u}$ is a well-defined homomorphism. Then construct a homomorphism $w : L \longrightarrow \mathbf{Z}/2\mathbf{Z}$ by first setting

$$w(\alpha_i) = \begin{cases} \overline{u}(\alpha_i) & \text{if } 1 \leq i \leq n \\ 0 & \text{if } n+1 \leq i \leq 2n \end{cases}$$

for the simple roots, and then extending w linearly to L. Let $\hat{\nu}$ be the antiautomorphism of $\hat{L}$ given by

$$\hat{\nu}(a) = \kappa^{w(\overline{a^{-1}})}\eta(a^{-1}) \tag{2.6.2}$$

for $a \in \hat{L}$. We first observe that $\overline{\hat{\nu}(a)} = \nu(\overline{a})$ for $a \in \hat{L}$. Then we see that for $a \in \hat{L}$ with $\overline{a} = \alpha_i$ and $1 \leq i \leq 2n$, we have $\hat{\nu}^2(a) = \kappa^{u(a)}\eta\left((\eta(a^{-1}))^{-1}\right)$. Since η fixes

κ we have $\left(\eta(a^{-1})\right)^{-1} = \eta(a)$. Thus $\eta\left(\left(\eta(a^{-1})\right)^{-1}\right) = \eta\left(\eta(a)\right) = \eta^2(a)$. So $\hat{\nu}^2(a) = \kappa^{u(a)}\left(\eta\right)^2(a) = \kappa^{2u(a)}a = a$, and $\hat{\nu}^2 = 1$, as required.

Using (2.6.2) we define an automorphism $\hat{\nu}$ on the finite dimensional Lie algebra $\underline{g}$ as follows:

$$\begin{aligned} \hat{\nu}(x_a) &= -x_{\hat{\nu}a} \text{ for } a \in \hat{L}, \\ \hat{\nu}|_{\underline{h}} &= \nu. \end{aligned} \tag{2.6.3}$$

Then extend $\hat{\nu}$ to an automorphism τ of $\tilde{\underline{g}}$ by

$$\tau(x \otimes t^m + rc + sd) = (-1)^m(\hat{\nu}x) \otimes t^m + rc + sd$$

where $x \in \underline{g}$, $m \in \mathbf{Z}$ and r, $s \in \mathbf{C}$. The fixed subalgebra $\tilde{\underline{g}}^{(\tau)}$ is given by $\amalg_{n\in\mathbf{Z}}(\tilde{\underline{g}}^{(\tau)}_{[n]} \otimes t^n) \oplus \mathbf{C}c \oplus \mathbf{C}d$. The form $\langle\cdot,\cdot\rangle$ on $\tilde{\underline{g}}$ also restricts to a $\tilde{\underline{g}}^{(\tau)}$-invariant form $\langle\cdot,\cdot\rangle$ on $\tilde{\underline{g}}^{(\tau)}$ and the fixed subalgebra $\tilde{\underline{g}}^{(\tau)}$ of $\tilde{\underline{g}}$ is isomorphic to $A_{2n}^{(2)}$ (cf. §3.5 in [M]).

Define $\Delta_{\pm 1} = \{\alpha \in \Delta \mid (-1)^{\langle \alpha, \nu\alpha\rangle} = \pm 1\}$. For $a \in \hat{L}$ such that $\overline{a} \in \Delta$, set

$$\overline{x}_a = \begin{cases} x_a & \text{if } \overline{a} \in \Delta_1 \\ \sqrt{2}x_a & \text{if } \overline{a} \in \Delta_{-1}. \end{cases} \tag{2.6.4}$$

Let $\Delta_+ = \{\alpha \in \Delta \mid \alpha = \sum_{i=1}^{2n} n_i\alpha_i,\ n_i \in \mathbf{N}\}$ denote the set of positive roots. As in earlier sections, let $\check{\Delta}_+$ be a complete set of orbit representatives of Δ_+ under the group generated by ν. Put $\check{\Delta} = \check{\Delta}_+ \cup (-\check{\Delta}_+)$.

Lemma 2.6.1 *Let $\alpha \in \Delta^1$.*

(1) $\alpha \in \Delta_1^1 \iff \langle \nu\alpha, \alpha\rangle = 0$.

(2) $\alpha \in \Delta_{-1}^1 \iff \langle \nu\alpha, \alpha\rangle = -1$.

Proof: Without loss of generality, we may assume α is a positive root. If $\alpha \in \Delta_1^1$, then $\langle \nu\alpha, \alpha\rangle = -2$ or 0. Since ν is a graph automorphism ν must preserve the positive roots; thus $\langle \nu\alpha, \alpha\rangle \neq -2$. Hence $\langle \nu\alpha, \alpha\rangle = 0$.

Next suppose that $\alpha \in \Delta_{-1}^1$. By definition we must have $\langle \nu\alpha, \alpha\rangle = \pm 1$. If $\langle \nu\alpha, \alpha\rangle = 1$, then $\nu\alpha - \alpha$ is a root. Thus we notice that $\nu\left(\nu\alpha - \alpha\right) = \alpha - \nu\alpha = -\left(\nu\alpha - \alpha\right)$. However this is impossible since ν preserves the positive roots. Hence $\langle \nu\alpha, \alpha\rangle = 1$. □

Using (2.1.7), (2.1.11), (2.1.12), (2.2.4), (2.6.3), (2.6.4) and Lemma 2.6.1 we have (cf. [M]):

Theorem 2.6.2 *The algebra* $\tilde{\underline{\mathbf{g}}}^{(\tau)}$ *is isomorphic to* $A_{2n}^{(2)}$. *Moreover* $\tilde{\underline{\mathbf{g}}}^{(\tau)}$ *has commutation relations given by:*

for $\alpha,\ \beta \in \check{\Delta}$, $b \in \hat{\check{\Delta}}$ *and* $m,\ n \in \mathbf{Z}$

$$[\alpha^+_{[m]}\otimes t^m, \beta^+_{[n]}\otimes t^n] = m\langle \alpha^+_{[m]}, \beta^+_{[n]}\rangle \delta_{m+n,\,0}c; \tag{2.6.5}$$

$$[\alpha^+_{[m]}\otimes t^m, \overline{x}^+_{b,\,[n]}\otimes t^n] = \langle \alpha^+_{[m]}, \overline{b}_{[m]}\rangle \overline{x}^+_{b,\,[m+n]}\otimes t^{m+n}; \tag{2.6.6}$$

for $a \in \hat{\check{\Delta}}$ *and* $m,\ n \in \mathbf{Z}$

$$[\overline{x}^+_{a,\,[m]}\otimes t^m, \overline{x}^+_{a^{-1},\,[n]}\otimes t^n]$$

$$= \begin{cases} \overline{a}^+_{[m+n]}\otimes t^{m+n} + m\langle \overline{x}^+_{a,\,[m]}, \overline{x}^+_{a^{-1},\,[n]}\rangle \delta_{m+n,\,0}c & \text{if } \overline{a} \in \Delta_1 \\ 2(\overline{a}^+_{[m+n]}\otimes t^{m+n} + m\langle \overline{x}^+_{a,\,[m]}, \overline{x}^+_{a^{-1},\,[n]}\rangle \delta_{m+n,\,0}c & \text{if } \overline{a} \in \Delta_{-1}; \end{cases} \tag{2.6.7}$$

for $a, b \in \hat{\check{\Delta}}$ *and* $m, n \in \mathbf{Z}$ *such that* $-1 \le \langle \overline{a}, \overline{b}\rangle \le \langle \nu\overline{a}, \overline{b}\rangle$ *and* $\overline{x}^+_{a,\,[m]} \neq 0 \neq \overline{x}^+_{b,\,[n]}$,

$$[\overline{x}^+_{a,\,[m]}\otimes t^m, \overline{x}^+_{b,\,[n]}\otimes t^n]$$

$$= \begin{cases} \overline{x}^+_{ab,\,[m+n]}\otimes t^{m+n} & \text{if } \overline{ab} \in \Delta^1 \text{ with } \overline{a} \text{ or } \overline{b} \in \Delta_1, \\ 2\overline{x}^+_{ab,\,[m+n]}\otimes t^{m+n} & \text{if } \langle \overline{a}, \overline{b}\rangle = -1 \text{ with either } \overline{ab} \in \Delta^1_1 \\ & \quad \text{and } \overline{a},\ \overline{b} \in \Delta_{-1}, \text{ or } \overline{ab} \in \Delta^0 \\ & \quad \text{with } \overline{a},\ \overline{b} \in \Delta^1_1 \\ 4\overline{x}^+_{ab,\,[m+n]}\otimes t^{m+n} & \text{if } \langle \overline{a}, \overline{b}\rangle = -1 \text{ with } \overline{ab} \in \Delta_0 \\ & \quad \text{and } \overline{a},\ \overline{b} \in \Delta_{-1} \\ 0 & \text{if } \langle \overline{a}, \overline{b}\rangle \ge 0; \end{cases} \tag{2.6.8}$$

for $x \in \underline{\mathbf{g}}$ *and* $m \in \mathbf{Z}$,

$$[d, \overline{x}^+_{[m]}\otimes t^m] = m\overline{x}^+_{[m]}\otimes t^m. \tag{2.6.9}$$

Remark 2.6.3 As in earlier sections, the above bracket relations may be written in terms of a section e satisfying (2.2.10) (cf. Remark 2.2.3).

Using Theorem 2.6.2 we can describe the Cartan subalgebra of $\tilde{\mathfrak{g}}^{(\tau)}$ by (2.2.11). Also by Theorem 2.6.2 we see that the set of roots for $\tilde{\mathfrak{g}}^{(\tau)}$ is given by

$$\begin{aligned}\Delta(\tilde{\mathfrak{g}}^{(\tau)}) &= \{\alpha_{[0]} + (2m+1)c \mid \alpha \in \Delta \text{ and } m \in \mathbf{Z}\} \cup \{\beta_{[0]} + 2mc \mid \beta \in \Delta^1 \text{ and } m \in \mathbf{Z}\} \\ &\quad \cup \{nc \mid n \in \mathbf{Z} \setminus \{0\}\} \\ &= \Delta_R(\tilde{\mathfrak{g}}^{(\tau)}) + \Delta_I(\tilde{\mathfrak{g}}^{(\tau)}).\end{aligned}$$

Choose a section e so that the associated 2-cocycle ϵ_0 satisfies (2.2.10). Define a *Chevalley basis* of $\tilde{\mathfrak{g}}^{(\tau)}$ [M] to be a basis for $\tilde{\mathfrak{g}}^{(\tau)}$ of the form

$$\begin{aligned}&\{\overline{x}^+_{\alpha+(2m+1)c}\ \left(\in (\tilde{\mathfrak{g}}^{(\tau)})^{\alpha_{[0]}+(2m+1)c}\right) \mid \alpha \in \Delta \text{ and } m \in \mathbf{Z}\} \\ &\cup \{\overline{x}^+_{\beta+(2m)c}\ \left(\in (\tilde{\mathfrak{g}}^{(\tau)})^{\beta_{[0]}+2mc}\right) \mid \beta \in \Delta^1 \text{ and } m \in \mathbf{Z}\} \\ &\cup \{\alpha^\vee_{i,\,[0]} \otimes t^{2m} \mid 1 \le i \le n-1\} \cup \{\alpha^\vee_{i,\,[1]} \otimes t^{2m+1} \mid 1 \le i \le n,\ m \in \mathbf{Z}\} \\ &\cup \{\alpha^\vee_{n,\,[0]} \otimes 1\} \cup \{\tfrac{1}{2}\alpha^\vee_{n,\,[0]} \otimes t^{2m} \mid m \in \mathbf{Z} \setminus \{0\}\} \\ &\cup \{(\gamma + c)^\vee\} \cup \{d\}\end{aligned}$$

where $\gamma = -\sum_{i=1}^{2n} \alpha_i$, such that conditions 1 and 2 of the definition in §2.3 hold.

Lemma 2.6.4 *1) If $\alpha \in \Delta^1_1 \cup \Delta^0$, then $\alpha^\vee_{[0]} = \alpha^+_{[0]}$.*
2) If $\alpha \in \Delta^1_{-1}$, then $\alpha^\vee_{[0]} = 2\alpha^+_{[0]}$.

Proof: The result is a consequence of the definition of "$\vee$" and Lemma 2.6.1. □

Lemma 2.6.4 implies that

$$(\alpha_{[0]} + mc)^\vee = \begin{cases} \alpha^\vee_{[0]} + 2mc & \text{if } \alpha \in \Delta^1_1 \\ \alpha^\vee_{[0]} + mc & \text{if } \alpha \in \Delta^0 \\ \alpha^\vee_{[0]} + 4mc & \text{if } \alpha \in \Delta^1_{-1}. \end{cases} \tag{2.6.10}$$

Thus as a consequence of Theorem 2.6.2 and (2.6.10) we have ([M]):

Theorem 2.6.5 *Choose a section e and corresponding 2-cocycle ϵ_0 which satisfies both (2.2.10) and $\epsilon_0 \equiv \frac{\langle \alpha, \alpha \rangle}{2}$ mod $2\mathbf{Z}$ for $\alpha \in L$.. Then the set*

$$\mathbf{S}^{(\tau)} = \{\overline{x}^+_{\alpha,\,[1]} \otimes t^{2m+1} \mid \alpha \in \check{\Delta} \textit{ and } m \in \mathbf{Z}\}$$

$$\cup \{\overline{x}^{+}_{\beta,[0]} \otimes t^{2m} \mid \beta \in \check{\Delta}^1 \textit{ and } m \in \mathbf{Z}\}$$
$$\cup \{\alpha^{+}_{i,[m]} \otimes t^{m} \mid 1 \leq i \leq n-1 \textit{ and } m \in \mathbf{Z}\}$$
$$\cup \{\alpha^{+}_{n,[m]} \otimes t^{m} \mid m \in \mathbf{Z} \setminus \{0\}\} \cup \{2\alpha^{+}_{n,[0]} \otimes 1\}$$
$$\cup \{\gamma^{+}_{[0]} + c\} \cup \{d\}$$

is a Chevalley basis of $\tilde{\mathfrak{g}}^{(\tau)}$ *and the involution* θ *is an isometry with respect to the* $\tilde{\mathfrak{g}}^{(\tau)}$*-invariant form* $\langle \cdot, \cdot \rangle$.

Chapter 3

The Main Theorem

3.1 Integral bases of the universal enveloping algebras of the affine Lie algebras

We begin this section with a series of preliminary lemmas and definitions.

Let $\mathbf{A}$ be an associative algebra over $\mathbf{C}$. As usual $\mathbf{A}[[\zeta]]$ will denote the ring of formal power series in ζ with coefficients in $\mathbf{A}$. Let $X, Y \in \mathbf{A}[[\zeta]]$. Define r_X, l_X, and $ad_X \in End(\mathbf{A}[[\zeta]])$ as follows:

$$\begin{aligned} r_X(Y) &= YX, \\ l_X(Y) &= XY, \text{ and} \\ ad_X(Y) &= (l_X - r_X)(Y) = XY - YX. \end{aligned} \tag{3.1.1}$$

If $X \in \mathbf{A}[[\zeta]]$ has constant coefficient equal to zero, i.e., $X \in \zeta\mathbf{A}[[\zeta]]$, then any power series in X is well-defined in $\mathbf{A}[[\zeta]]$. Thus elements such as

$$\begin{aligned} \exp X &= \sum_{n\geq 0} \frac{X^n}{n!}, \\ \log(1+X) &= X - \frac{X^2}{2} + \frac{X^3}{3} - \cdots = \sum_{n\geq 1} \frac{(-1)^{n+1}X^n}{n} \quad \text{and} \\ (1-X)^{-1} &= \sum_{n\geq 0} X^n \end{aligned}$$

are all well-defined elements in $\mathbf{A}[[\zeta]]$ as long as $X \in \zeta\mathbf{A}[[\zeta]]$. From the Baker-Campbell-Hausdorff formula, we see that if X and Y are elements of $\zeta\mathbf{A}[[\zeta]]$ such that $[X,Y]=0$, i.e., $ad_X(Y)=0$, then

$$\exp X \exp Y = \exp(X+Y) = \exp Y \exp X. \tag{3.1.2}$$

Later in this section we will want to "reverse the order" of $\exp X \exp Y$ in cases where $[X,Y]$ may not equal zero. The following lemmas will be useful. For more details see §4.1 of [M].

Lemma 3.1.1 *Let* $X \in \zeta\mathbf{A}[[\zeta]]$. *Then* $l_{\exp X} = r_{\exp X} \cdot \exp ad_X$.

Lemma 3.1.2 *Let* $X \in \zeta\mathbf{A}[[\zeta]]$ *and* $D \in \zeta(Der\ \mathbf{A})[[\zeta]]$ $(\subset \zeta\ Der(\mathbf{A}[[\zeta]]))$. *Then*

$$(\exp D)(\exp X) = \exp((\exp D)(X)).$$

Lemma 3.1.3 *Let* X *and* Y *be in* $\zeta\mathbf{A}[[\zeta]]$ *and suppose that* $[X,[X,Y]]=[Y,[X,Y]]=0$. *Then*

$$\exp X \exp Y = \exp Y \exp\,[X,Y] \exp X.$$

Proof: Apply Lemma 3.1.1, Lemma 3.1.2 and (3.1.2). □

Let X_1, X_2, ..., $X_n, \ldots$ be independent variables and let $\mathbf{C}[X_1, X_2, \ldots]$ denote the algebra of polynomials in the X_i with coefficients in $\mathbf{C}$. For a nonnegative integer s, define $\Lambda_s(X) = \Lambda_s(X_1, X_2, \ldots, X_s)$ to be the coefficient of ζ^s in

$$\sum_{s\geq 0} \Lambda_s \zeta^s = \exp\left(\sum_{j\geq 1} \frac{X_j}{j}\zeta^j\right). \tag{3.1.3}$$

We note that the polynomials Λ_s are examples of a more general family of polynomials called Schur polynomials (cf. [M], [L-P], [Mac], and [F-L-M]). The Schur polynomials play an important role in the description of integral forms of vertex operator algebras (cf. [B]). We will discuss Schur polynomials in more detail in the later sections.

Define $\binom{X}{n} = \frac{X(X-1)\cdots(X-n+1)}{n!}$ for $X \in \mathbf{A}$ and $n \in \mathbf{N}$.

Lemma 3.1.4 *For* X *an indeterminate and* s *a nonnegative integer,*

$$\Lambda_s((X)) = \Lambda_s(X, X, \ldots, X) = \binom{X+s-1}{s}.$$

Proof: For $s \geq 1$ and $X \in \mathbf{A}$ we have

$$\binom{X}{s} = \frac{X}{s}\binom{X-1}{s-1}. \tag{3.1.4}$$

Also for $s = 0$ we see that

$$\Lambda_0(X) = 1 = \binom{X-1}{0}. \tag{3.1.5}$$

Now let $s \geq 1$. By induction we have

$$\Lambda_{s-1}(X) = \binom{X+s-2}{s-1} = \frac{s}{X+s-1}\binom{X+s-1}{s}.$$

We next need to compare Λ_{s-1} and Λ_s. Take $\zeta\frac{\partial}{\partial\zeta}$ on both sides of the equation (3.1.3). Since $\frac{\partial}{\partial\zeta}(-X(1-\zeta))$ commutes with $-X(1-\zeta)$, we have

$$\frac{\partial}{\partial\zeta}\exp\left(-X\log(1-\zeta)\right) = X(1-\zeta)^{-1}\exp\left(-X\log(1-\zeta)\right).$$

Thus

$$\begin{aligned}
\sum_{s\geq 1} s\Lambda_s\zeta^s &= X\frac{\zeta}{1-\zeta}\exp(-X\log(1-\zeta)) \\
&= X\zeta(1+\zeta+\zeta^2+\cdots)(\exp(-X\log(1-\zeta))) \\
&= X\zeta(\exp(-X\log(1-\zeta))) + X\zeta^2(1-\zeta)^{-1}\exp(-X\log(1-\zeta)) \\
&= \sum_{s\geq 1} X\Lambda_{s-1}((X))\zeta^s + \zeta^2\frac{\partial}{\partial\zeta}(\exp(-X\log(1-\zeta))) \\
&= \sum_{s\geq 1} X\Lambda_{s-1}((X))\zeta^s + \sum_{s\geq 1}(s-1)\Lambda_{s-1}((X))\zeta^s.
\end{aligned}$$

So for $s \geq 1$,we have $s\Lambda_s = (X+s-1)\Lambda_{s-1}$. Hence we obtain

$$\frac{X+s-1}{s}\Lambda_{s-1}(X) = \Lambda_s = \binom{X+s-1}{s}. \square$$

Let $\underline{g}$ be a Lie algebra as in §2.1. Next let $f(\zeta_1,\zeta_2) = \sum_{r,s\geq 0} a_{r,s}\zeta_1^r\zeta_2^s$ be an element of $\mathbf{C}[[\zeta_1,\zeta_2]]$. Then for $u \in \underline{g}$ and $m,\ n \in \mathbf{Z}$ set

$$u(f(t^m\zeta_1, t^n\zeta_2)) = \sum_{r,s\geq 0} a_{r,s}(u\otimes t^{mr+ns})\zeta_1^r\zeta_2^s. \tag{3.1.6}$$

We remark that we will generally use the one variable analogue of (3.1.6).

The following theorem will play an important role in the straightening arguments of future sections.

Theorem 3.1.5 *Let u and v be elements of $\underline{\mathrm{g}}$. Then for $m, n \in \mathbf{Z}$ and $f(\zeta), g(\zeta) \in \zeta\mathbf{C}[[\zeta]]$ we have*

$$\begin{aligned}&\exp(v(f(t^n\zeta_2)))\cdot\exp(u(g(t^m\zeta_1))) = \\ &\quad\exp\left(\sum_{k\geq 0}\frac{(ad\, v)^k\cdot u)}{k!}\left((f(t^n\zeta_2))^k\cdot g(t^m\zeta_1)\right)\right)\times\\ &\quad\times\exp\left(\sum_{k\geq 1}\frac{\langle v,(ad\, v)^{k-1}\cdot u)\rangle}{k!}\cdot \mathrm{Res}_t\left[\frac{\partial f(t^n\zeta_2)}{\partial t}\cdot(f(t^n\zeta_2))^{k-1}\cdot g(t^m\zeta_1)\right]c\right)\times\\ &\quad\times\exp(v(f(t^n\zeta_2))).\end{aligned}$$

Proof: By Lemma 3.1.1 and Lemma 3.1.2

$$\begin{aligned}&\exp(v(f(t^n\zeta_2)))\cdot\exp(u(g(t^m\zeta_1))) = \\ &\quad\exp((\exp \mathrm{ad}\;\; v(f(t^n\zeta_2)))\cdot u(g(t^m\zeta_1)))\cdot\exp(v(f(t^n\zeta_2))).\end{aligned}$$

We shall apply an induction argument to see that

$$\begin{aligned}\frac{(\mathrm{ad}\;\; v(f(t^n\zeta_2)))^k}{k!}\cdot u(g(t^m\zeta_1)) = {}& \frac{v_0^k\cdot u}{k!}((f(t^n\zeta_2))^k\cdot g(t^m\zeta_1))+ \\ &\left(\frac{\langle v,(\mathrm{ad}\; v)^{k-1}\cdot u\rangle}{k!}\right)\left(\mathrm{Res}_t\left[\frac{\partial(f(t^n\zeta_2))}{\partial t}((f(t^n\zeta_2))^{k-1}\cdot g(t^m\zeta_1))\right]\right)\end{aligned}\tag{3.1.7}$$

for $k \geq 1$.

By (2.1.11), (2.1.12) and (3.1.6)

$$\begin{aligned}[v(t^{ns}\zeta_2^s), u(t^{mr}\zeta_1^r)] &= ((\mathrm{ad}\; v)\cdot u)(t^{ns+mr}\zeta_1^r\zeta_2^s) + ns\langle v,u\rangle\delta_{ns+mr,\,0}\zeta_1^r\zeta_2^s c \\ &= ((\mathrm{ad}\; v)\cdot u)(t^{ns+mr}\zeta_1^r\zeta_2^s) + \langle v,u\rangle\left[\mathrm{Res}_t\frac{\partial(t^{ns}\zeta_2^s)}{\partial t}t^{mr}\zeta_1^r\right]c.\end{aligned}$$

Thus

$$\begin{aligned}&ad\, v(f(t^n\zeta_2))\cdot u(g(t^m\zeta_1)) = \\ &\quad((\mathrm{ad}\; v)\cdot u)(f(t^n\zeta_2)\cdot g(t^m\zeta_1)) + \langle v,u\rangle\left[\mathrm{Res}_t\frac{\partial(f(t^n\zeta_2))}{\partial t}\cdot g(t^m\zeta_1)\right]c;\end{aligned}$$

and so the result is true for $k = 1$.

Next suppose for $k \geq 1$ that (3.1.7) is true. Then for $k+1$, using (2.1.11) we have

$$\frac{(\mathrm{ad}\; v(f(t^n\zeta_2)))^{k+1}}{(k+1)!}\cdot u(g(t^m\zeta_1)) =$$

$$= \left(\frac{1}{k+1}\right)(\text{ad } v\,(f(t^n\zeta_2)))\left\{\frac{((\text{ad } v)^k\cdot u)}{k!}\left((f(t^n\zeta_2))^k\cdot g(t^m\zeta_1)\right)\right.$$
$$\left.+\frac{\langle v,(\text{ad } v)^{k-1}\cdot u\rangle}{k!}\left[\text{Res}_t\frac{\partial(f(t^n\zeta_2))}{\partial t}((f(t^n\zeta_2))^{k-1}\cdot g(t^m\zeta_1))\right]c\right\}$$
$$= \frac{((\text{ad } v)^{k+1}\cdot u)}{(k+1)!}\left((f(t^n\zeta_2))^{k+1}\cdot g(t^m\zeta_1)\right)$$
$$+\frac{\langle v,(\text{ad } v)^{k}\cdot u\rangle}{(k+1)!}\left[\text{Res}_t\frac{\partial(f(t^n\zeta_2))}{\partial t}\left((f(t^n\zeta_2))^{k}\cdot g(t^m\zeta_1)\right)\right]c.$$

Then induction and (3.1.2) imply

$$\exp\left(\exp(\text{ad } \ v(f(t^n\zeta_2)))\cdot u(g(t^m\zeta_1))\right) =$$
$$= \exp\left(\sum_{k\geq 0}\frac{((\text{ad } v)^k\cdot u)}{k!}\left((f(t^n\zeta_2))^k\cdot g(t^m\zeta_1)\right)\right)$$
$$\times\exp\left(\sum_{k\geq 1}\frac{\langle v,(\text{ad } v)^{k-1}\cdot u\rangle}{k!}\left[\text{Res}_t\frac{\partial(f(t^n\zeta_2))}{\partial t}\left((f(t^n\zeta_2))^{k-1}\cdot g(t^m\zeta_1)\right)\right]c\right).\ \square$$

We next present the theorem in Chapter 4 of [M] which gives an explicit description of the integral basis (and hence its **Z**-span or integral form) of the universal enveloping algebra associated to the affine algebras $\tilde{\underline{\mathbf{g}}}$, $\tilde{\underline{\mathbf{g}}}_{[0]}$ and $\tilde{\underline{\mathbf{g}}}^{(\tau)}$. Our treatment differs in that we first derive the exponential identities needed to prove the theorem for the simply-laced affines $\tilde{\underline{\mathbf{g}}}$. Then we realize the identities required to prove the straightening arguments needed for the remaining affines $\tilde{\underline{\mathbf{g}}}_{[0]}$ and $\tilde{\underline{\mathbf{g}}}^{(\tau)}$, as consequences of the identities for $\tilde{\underline{\mathbf{g}}}$. The proofs offered here actually construct the identities instead of merely checking their validity (cf. [M])

In order to give a concise statement of the theorem, we must introduce some notation (cf. [M]). For brevity, let $\underline{\ell}$ denote any of the affines listed in the above paragraph, $\Pi = \{\alpha_{i,\,[0]}\mid i\in I \text{ or } J\}$ and let $\gamma\in\Delta$ be the lowest weight of the $\underline{\mathbf{g}}_{[0]}$-module $\underline{\mathbf{g}}_{[\pm 1]}$. Note that if $\underline{\ell} = \tilde{\underline{\mathbf{g}}}$, then $\nu = id$, $\gamma = \alpha_0$ and $I = \{1, 2, \ldots, l\}$. Set

$$X_{0,\,0} \;=\; (\gamma_{[0]} + c)^{\vee}, \tag{3.1.8}$$

$$X_{n,i} = \begin{cases} \frac{1}{2}(\alpha_{i,[0]}^{\vee}\otimes t^n) & \text{if } \alpha_i \in \Delta_{-1} \text{ and } n \in 2\mathbf{Z}\setminus\{0\} \\ (\alpha_{i,[0]}^{\vee}\otimes t^n) & \text{if } n \neq 0 \text{ and } [n]=[0] \\ (\alpha_{i,[0]}^{\vee}\otimes 1) & \text{if } n=0 \\ (\alpha_{i,[0]}^{+}\otimes t^n) & \text{if } [n]\neq[0] \text{ and } \Delta = \check{\Delta}(\underline{\ell}^{(\tau)}), \end{cases} \tag{3.1.9}$$

$$\bar{I} = \{(n,i) \in \mathbf{Z}\times I \mid X_{n,i} \neq 0\}, \text{ and} \tag{3.1.10}$$

$$\hat{I} = \bar{I} \cup \{(0,0)\}. \tag{3.1.11}$$

Let $\mathbf{S}$ be a Chevalley basis of $\underline{\ell}$. We notice that we may replace the imaginary root vectors and the Cartan subalgebra elements found in the descriptions of the Chevalley bases of §§2.1-2.6 with the $X_{n,i}$'s defined in (3.1.8) and (3.1.9). Fix a linear order on $\mathbf{S}$. Choose nonnegative integers $s(m,\overline{a})$ and $s(n,i)$ for each $\overline{a}_{[0]}(-1)_0 + mc \in \Delta$ and each $(n,i) \in \hat{I}$ such that finitely many are nonzero. Then for an integer $s \geq 0$, define the monomial associated to $s(m,\overline{a})$, $s(n,i)$, s to be the (ordered) product of the elements

$$\Lambda_{s(m,\overline{a})}\left((\overline{x}^{+}_{a^N,[Nm]}\otimes t^{Nm})_N\right) = \Lambda_{s(m,\overline{a})}(\overline{x}^{+}_{a,[m]}\otimes t^m, 0, \ldots, 0) = \frac{(\overline{x}^{+}_{a,[m]}\otimes t^m)^{s(m,\overline{a})}}{s(m,\overline{a})!}; \tag{3.1.12}$$

$$\Lambda_{s(n,i)}\left((X_{Nn,i})_N\right) = \Lambda_{s(n,i)}\left(X_{n,i}, X_{2n,i}, \ldots, X_{s(n,i)n,i}\right); \tag{3.1.13}$$

$$\Lambda_s((d)) = \binom{d+s-1}{s}. \tag{3.1.14}$$

Theorem 3.1.6 ([G], [M]) *Fix a linearly ordered Chevalley basis* $\mathbf{S}$ *of* $\underline{\ell}$. *Let* $U_{\mathbf{Z}} = U_{\mathbf{Z}}(\underline{\ell})$ *be the* $\mathbf{Z}$*-subalgebra of* $U(\underline{\ell})$ *generated by the elements (3.1.12) and (3.1.14) where* $\overline{a}_{[0]} + mc \in \Delta$ *and* $s \in \mathbf{N}$. *Then the set of (ordered) monomials forms an integral basis of* $U_{\mathbf{Z}}$, *and* $U_{\mathbf{Z}}$ *is an integral form of* $U(\underline{\ell})$, *i.e.,* $U_{\mathbf{Z}} \otimes_{\mathbf{Z}} \mathbf{C} = U(\underline{\ell})$.

Proof: Since the main framework of the proof we present is that found in [M], we will only give a sketch of the argument used in [M] below. We remind the reader that our argument will differ from [M] in the manner in which the identities needed to prove the essential straightening arguments are verified.

Let $U'_{\mathbf{Z}}$ denote the $\mathbf{Z}$-subalgebra of $U(\underline{\ell})$ generated by the set of (ordered) monomials. To prove the theorem we must demonstrate the following:

i) $U'_{\mathbf{Z}} = U_{\mathbf{Z}}$,

ii) the set of (ordered) monomials forms a $\mathbf{Z}$-basis of $U'_{\mathbf{Z}}$, i.e., any product of monomials can be written as a $\mathbf{Z}$-linear combination of (ordered) monomials.

The Poincaré-Birkhoff-Witt Theorem implies the set of (ordered) monomials forms a basis of $U(\underline{\ell})$ over $\mathbf{C}$, so once we have shown i) and ii) to be true, we have $U_{\mathbf{Z}} \otimes_{\mathbf{Z}} \mathbf{C} = U(\underline{\ell})$.

Clearly we have $U_{\mathbf{Z}} \subset U'_{\mathbf{Z}}$. To get the other inclusion, we will have to show that for each $(n, i) \in \hat{I}$ and $s \geq 0$, $\Lambda_s((X_{Nn,i})_N) \in U_{\mathbf{Z}}$.

To prove ii) we will use straightening arguments. These type of arguments require that we define the notion of *degree* on $U(\underline{\ell})$. Let

$$\mathbf{C} = U_{(0)}(\underline{\ell}) \subset U_{(1)}(\underline{\ell}) \subset U_{(2)}(\underline{\ell}) \subset \cdots$$

denote the usual filtration of $U(\underline{\ell})$. Then for $u \in U(\underline{\ell})$, define the *degree of* u to be k if $u \in U_{(k)}(\underline{\ell})$ with $u \notin U_{(k-1)}(\underline{\ell})$.

The straightening arguments will show

1. for any two products of monomials C_1 and C_2 containing factors of the form (3.1.12), (3.1.13) or (3.1.14), we have

$$C_1 C_2 = C_2 C_1 + P$$

where P is a $\mathbf{Z}$-linear combination of products of monomials Q such that $deg\, Q < deg\, C_1 + deg\, C_2$; and

2. for r, $s \geq 0$, we have

$$\Lambda_r((Y_N)_N)\Lambda_s((Y_N)_N) = \binom{r+s}{s} \Lambda_{r+s}((Y_N)_N) + P$$

where P is a **Z**-linear combination of products Q of monomials such that $degQ < r+s$, and where $N \geq 1$, Y_N is one of $\overline{x}^+_{a,\,[Nm]}\otimes t^{Nm}$ (with $\overline{a}_{[0]} + mc \in \Delta$), $X_{Nn,\,i}$ (with $(n,i) \in \hat{I}$) or d.

To prove the first assertion it is clearly sufficient to prove it in the case that C_1 and C_2 have the form (3.1.12), (3.1.13) or (3.1.14). We must consider the following possible choices for C_1 and C_2:

Case 1:

$C_1 = \Lambda_r\left((\overline{x}^+_{a,\,[Nm]}\otimes t^{Nm})_N\right)$ and $C_2 = \Lambda_s\left((\overline{x}^+_{b,\,[Nn]}\otimes t^{Nn})_N\right)$ with $\overline{a}_{[0]}$, $\overline{b}_{[0]} \in \Delta$ and $\overline{ab}_{[0]} \notin \Delta \cup \{0\}$;

Case 2:

$C_1 = \Lambda_r\left((X_{Nm,\,i})_N\right)$ and $C_2 = \Lambda_s\left((X_{Nn,\,j})_N\right)$ with (m,i), $(n,j) \in \hat{I}$ and $mn \geq 0$;

Case 3:

$C_1 = \Lambda_r((d))$ and $C_2 = \Lambda_s((d))$;

Case 4:

$C_1 = \Lambda_r\left((\overline{x}^+_{a,\,[Nm]}\otimes t^{Nm})_N\right)$ and $C_2 = \Lambda_s\left((\overline{x}^+_{b,\,[Nn]}\otimes t^{Nn})_N\right)$ with $\overline{a}_{[0]}$, $\overline{b}_{[0]} \in \Delta$ and $\overline{ab}_{[0]} \in \Delta \cup \{0\}$;

Case 5:

$C_1 = \Lambda_r\left((\overline{x}^+_{a,\,[Nm]}\otimes t^{Nm})_N\right)$ and $C_2 = \Lambda_s\left((X_{Nn,i})_N\right)$ for $\overline{a}_{[0]} \in \Delta$ and $(n,i) \in \hat{I}$;

Case 6:

$C_1 = \Lambda_r\left((X_{Nm,i})_N\right)$ and $C_2 = \Lambda_s\left((X_{-Nn,j})_N\right)$ for (m,i) and $(n,j) \in \hat{I}$ $m,\, n > 0$;

Case 7:

$C_1 = \Lambda_r\left((\overline{x}^+_{a,\,[Nm]}\otimes t^{Nm})_N\right)$ and $C_2 = \Lambda_s((d))$ for $\overline{a}_{[0]} \in \Delta$;

Case 8:

$C_1 = \Lambda_r\left((X_{Nn,i})_N\right)$ and $C_2 = \Lambda_s((d))$ with $(n,i) \in \hat{I}$.

It is not hard to see in the first three cases that $C_1C_2 = C_2C_1$. For the remaining five possibilities, we first observe that the monomials (3.1.12), (3.1.13), and (3.1.14), respectively, occur as the coefficients of $\zeta^{s(m,\overline{a})}$, $\zeta^{s(n,i)}$ and ζ^s, respectively, in the exponential generating functions

$$\exp\left(\overline{x}^{+}_{a,\,[Nm]}\otimes t^{Nm}\zeta\right),\ \exp\left(\sum_{j\geq 1}\frac{X_{jn,i}}{j}\zeta^{j}\right)\ \text{and}\ \exp\left(\sum_{j\geq 1}\frac{d}{j}\zeta^{j}\right),$$

respectively. (In the last two cases this follows from (3.1.3) and Lemma 3.1.4.) We prove identities which "reverse" the products of certain pairs of the above exponential generating functions. The commutation formulas for the products C_1C_2 are then read off as the homogeneous components of the exponential identities. One finds that the left hand side of these identities is just C_1C_2, while the terms on the right hand side have C_2C_1 as the highest degree term, plus a **Z**-linear combination of lower degree products of monomials of type (3.1.12), (3.1.13) and (3.1.14). These remaining cases will take more work and will be proven in the succeeding sections.

To see that assertion (2) holds, first observe that

$$\begin{aligned}
&\exp\left(\sum_{j\geq 1}\frac{Y_j}{j}{\zeta_1}^{j}\right)\exp\left(\sum_{j\geq 1}\frac{Y_j}{j}{\zeta_2}^{j}\right)=\\
&=\exp\left(\sum_{j\geq 1}\frac{Y_j}{j}({\zeta_1}^{j}+{\zeta_2}^{j})\right)\\
&=\exp\left(\sum_{j\geq 1}\frac{Y_j}{j}\left[(\zeta_1+\zeta_2)^j-\sum_{k=1}^{j-1}\binom{j}{k}{\zeta_1}^{j-k}{\zeta_2}^{k}\right]\right) \qquad (3.1.15)\\
&=\exp\left(\sum_{j\geq 1}\frac{Y_j}{j}(\zeta_1+\zeta_2)^j\right)\exp\left(-\sum_{j,\,k\geq 1}\binom{j+k}{j}\frac{Y_{j+k}}{j+k}{\zeta_1}^{j}{\zeta_2}^{k}\right).
\end{aligned}$$

Next recursively define the function $f_{a,\,b}(n)$ for $n\geq 1$ by

$$\frac{1}{a+b}\binom{an+bn}{an}=\sum_{\substack{d\geq 1,\\ d|n}}f_{a,\,b}(d)d$$

where $\gcd(a,b)=1$ and $a,\ b\geq 1$. Then we have

$$\begin{aligned}
&\exp\left(-\sum_{j,k\geq 1}\frac{1}{j+k}\binom{j+k}{j}Y_{j+k}{\zeta_1}^{j}{\zeta_2}^{k}\right)=\\
&=\exp\left(-\sum_{\substack{a,b\geq 1\\ \gcd(a,b)=1}}\left(\sum_{n\geq 1}f_{a,b}(n)\sum_{j\geq 1}\frac{Y_{(an+bn)j}}{j}({\zeta_1}^{an}{\zeta_2}^{bn})^j\right)\right)\\
&=\prod_{\substack{a,\,b\geq 1\\ \gcd(a,b)=1}}\left[\prod_{n\geq 1}\exp\left(-f_{a,\,b}(n)\sum_{j\geq 1}\frac{Y_{(an+bn)j}}{j}({\zeta_1}^{an}{\zeta_2}^{bn})^j\right)\right].
\end{aligned}$$

An induction argument on n shows that $f_{a,b}(n) \in \mathbf{Z}$. Thus we see that the second exponential generating function in (3.1.15) is a product of integral multiples of allowable monomials of smaller degree.

Thus Theorem 3.1.6 is proved for Cases 1-3.

3.2 Exponential identities for the simply-laced affine Lie algebras

In this section we will derive the exponential identities needed to prove Theorem 3.1.6 for the type 1 affines associated to the root lattices of type A, D or E. These identities will also be used in the proof of exponential identities for the other affine algebras (cf. Sections 3.3 - 3.7).

Proposition 3.2.1 *Let $u = x_a$ and $v = x_b$ where $\bar{a}$, $\bar{b} \in \Delta$. If $\langle \bar{a}, \bar{b} \rangle = -1$ then for m, $n \in \mathbf{Z}$ we have*

$$\exp(v(t^n\zeta_2))\exp(u(t^m\zeta_1)) = \exp(u(t^m\zeta_1))\exp(-[u,v](t^{m+n}\zeta_1\zeta_2))\exp(v(t^n\zeta_2)). \tag{3.2.1}$$

Proof: For the above choice of u and v, we see that $\langle v, u \rangle = 0$, $\langle v, (\mathrm{ad}\ v)^{k-1} \cdot u \rangle = 0$ for $k \geq 1$, and $(\mathrm{ad}\ v)^k \cdot u = 0$ for $k \geq 2$. After applying Theorem 3.1.5, the left hand side of (3.2.1) becomes

$$\exp(u(t^m\zeta_1) + [v,u](t^{m+n}\zeta_1\zeta_2))\exp(v(t^n\zeta_2)).$$

Since $[[v,u],u] = 0 = [u,[v,u]]$, Lemma 3.1.3 yields the result. □

We next prove a generalization of Lemma 4.3.4 (ii) in [M].

Proposition 3.2.2 *Let $u = x_a$ and $v = x_b$ with $\bar{a}$, $\bar{b} \in \Delta$ and $\langle \bar{a}, \bar{b} \rangle = -2$. Then for $f(\zeta)$, $g(\zeta) \in \mathbf{A}[[\zeta]]$ and m, n, $k \in \mathbf{Z}$, $k \geq 0$, we have*

$$\exp\left(v(t^{kn}\zeta_2^k f(t^{m+n}\zeta_1\zeta_2))\right)\exp\left(u(t^{km}\zeta_1^k g(t^{m+n}\zeta_1\zeta_2))\right) = \tag{3.2.2}$$

$$\exp\left(u(t^{km}\zeta_1^k g(t^{m+n}\zeta_1\zeta_2)(1+t^{km+kn}\zeta_1^k\zeta_2^k f(t^{m+n}\zeta_1\zeta_2)g(t^{m+n}\zeta_1\zeta_2))^{-1})\right)\times$$
$$\exp\left([v,u](\log(1+t^{km+kn}\zeta_1^k\zeta_2^k f(t^{m+n}\zeta_1\zeta_2)g(t^{m+n}\zeta_1\zeta_2)))\right)\times$$
$$\exp\left(\langle v,u\rangle kn\delta_{m+n,0}\log(1+\zeta_1^k\zeta_2^k f(\zeta_1\zeta_2)g(\zeta_1\zeta_2))c\right)\times$$
$$\exp\left(v(t^{kn}\zeta_2^k f(t^{m+n}\zeta_1\zeta_2)(1+t^{km+kn}\zeta_1^k\zeta_2^k f(t^{m+n}\zeta_1\zeta_2)g(t^{m+n}\zeta_1\zeta_2))^{-1})\right).$$

Proof: In this case we have

$$\begin{aligned}\exp(\mathrm{ad}\ v(t^{kn}\zeta_2^k f(t^{m+n}\zeta_1\zeta_2)))\cdot u(t^{km}\zeta_1^k g(t^{m+n}\zeta_1\zeta_2)) =\\ = u(t^{km}\zeta_1^k g(t^{m+n}\zeta_1\zeta_2)) + [v,u](t^{km+kn}\zeta_1^k\zeta_2^k f(t^{m+n}\zeta_1\zeta_2)g(t^{m+n}\zeta_1\zeta_2))\\ +\langle v,u\rangle kn\delta_{m+n,0}\zeta_1^k\zeta_2^k f(\zeta_1\zeta_2)g(\zeta_1\zeta_2)c\\ +\frac{((\mathrm{ad}\ v)^2\cdot u)}{2!}(t^{km+2kn}\zeta_1^k\zeta_2^{2k}(f(t^{m+n}\zeta_1\zeta_2))^2 g(t^{m+n}\zeta_1\zeta_2)).\end{aligned}$$

(Note: $\frac{(\mathrm{ad}\ v)^2\cdot u}{2!} = -v$.) Thus by Lemmas 3.1.1 and 3.1.2, the left hand side of (3.2.2) is given by

$$\begin{aligned}\exp\Big(u(t^{km}\zeta_1^k g(t^{m+n}\zeta_1\zeta_2)) + [v,u](t^{km+kn}\zeta_1^k\zeta_2^k f(t^{m+n}\zeta_1\zeta_2)g(t^{m+n}\zeta_1\zeta_2))\\ +\langle v,u\rangle kn\delta_{m+n,0}\zeta_1^k\zeta_2^k f(\zeta_1\zeta_2)g(\zeta_1\zeta_2)c\\ -v(t^{km+2kn}\zeta_1^k\zeta_2^{2k}(f(t^{m+n}\zeta_1\zeta_2))^2 g(t^{m+n}\zeta_1\zeta_2))\Big)\times\\ \times\exp(v(t^{kn}\zeta_2^k f(t^{m+n}\zeta_1\zeta_2))).\end{aligned}\tag{3.2.3}$$

To see that (3.2.3) is the right hand side in (3.2.2), we must prove the identity

$$\begin{aligned}\exp\Big(u(t^{km}\zeta_1^k g(t^{m+n}\zeta_1\zeta_2)) + [v,u]\left(t^{km+kn}\zeta_1^k\zeta_2^k f(t^{m+n}\zeta_1\zeta_2)g(t^{m+n}\zeta_1\zeta_2)\right.\\ \left.+\langle v,u\rangle kn\delta_{m+n,0}\zeta_1^k\zeta_2^k f(\zeta_1\zeta_2)g(\zeta_1\zeta_2)\right) c\Big)\\ -v(t^{km+2kn}\zeta_1^k\zeta_2^{2k}(f(t^{m+n}\zeta_1\zeta_2))^2 g(t^{m+n}\zeta_1\zeta_2))\Big)\\ = \exp\left(u(t^{km}\zeta_1^k g(t^{m+n}\zeta_1\zeta_2)(1+t^{km+kn}\zeta_1^k\zeta_2^k f(t^{m+n}\zeta_1\zeta_2)g(t^{m+n}\zeta_1\zeta_2))^{-1})\right)\times\\ \times\exp\Big([v,u](\log(1+t^{km+kn}\zeta_1^k\zeta_2^k f(t^{m+n}\zeta_1\zeta_2)g(t^{m+n}\zeta_1\zeta_2)))\\ +\langle v,u\rangle kn\delta_{m+n,0}\log(1+\zeta_1^k\zeta_2^k f(\zeta_1\zeta_2)g(\zeta_1\zeta_2))c\Big)\\ \times\exp\Big(-v(t^{km+2kn}\zeta_1^k\zeta_2^{2k}(f(t^{m+n}\zeta_1\zeta_2))^2 g(t^{m+n}\zeta_1\zeta_2)\\ \times(1+t^{km+kn}\zeta_1^k\zeta_2^k f(t^{m+n}\zeta_1\zeta_2)g(t^{m+n}\zeta_1\zeta_2))^{-1})\Big).\end{aligned}\tag{3.2.4}$$

To condense the computations which are needed to prove (3.2.4), we define the following:

$$\begin{aligned}
\overline{D} &= \frac{1}{k}\left(\zeta_1\frac{\partial}{\partial\zeta_1}+\zeta_2\frac{\partial}{\partial\zeta_2}\right) && (3.2.5)\\
f &= f(t^{m+n}\zeta_1\zeta_2) \\
g &= g(t^{m+n}\zeta_1\zeta_2) \\
p &= (1+t^{km+kn}{\zeta_1}^k{\zeta_2}^k fg)^{-1} && (3.2.6)\\
\phi_4 &= u(t^{km}{\zeta_1}^k g)+[v,u](t^{km+kn}{\zeta_1}^k{\zeta_2}^k fg)+ && (3.2.7)\\
&\qquad \langle v,u\rangle kn\delta_{m+n,\,0}{\zeta_1}^k{\zeta_2}^k fgc - v(t^{km+2kn}{\zeta_1}^k{\zeta_2}^{2k}(f)^2 g) \\
\phi_3 &= -u(t^{km}{\zeta_1}^k gp) && (3.2.8)\\
\phi_2 &= -[v,u](\log p^{-1}) - \langle v,u\rangle kn\delta_{m+n,\,0}\log(p^{-1})c && (3.2.9)\\
\phi_1 &= v(t^{km+2kn}{\zeta_1}^k{\zeta_2}^{2k}(f)^2 gp) && (3.2.10)\\
\Phi_i &= \exp\phi_i \quad \text{for } i=1,2,3,4 && (3.2.11)\\
\Psi &= \Phi_1\Phi_2\Phi_3\Phi_4. && (3.2.12)
\end{aligned}$$

The identity (3.2.4) will hold if we can demonstrate that $\Psi = 1$. We accomplish this by computing $\overline{D}\Psi$ and showing that $\overline{D}\Psi = 0$.

Using the product rule we have

$$\overline{D}\Psi = (\overline{D}\Phi_1)\Phi_2\Phi_3\Phi_4 + \cdots + \Phi_1\Phi_2\Phi_3(\overline{D}\Phi_4). \qquad (3.2.13)$$

We would like to replace (3.2.13) with

$$\left(\overline{D}\phi_1 + \exp\operatorname{ad}\,\phi_1\left(\overline{D}\phi_2 + \exp\operatorname{ad}\,\phi_2\left(\overline{D}\phi_3 + \exp\operatorname{ad}\,\phi_3\left(\overline{D}\phi_4\right)\right)\right)\right)\Psi.$$

However, this is not correct since $\overline{D}\Phi_4 \neq (\overline{D}\phi_4)\Phi_4$. To determine $\overline{D}\Phi_4$ we rewrite

$$\Phi_4 = \exp\left(\exp\operatorname{ad}\ v(t^{kn}{\zeta_2}^k f)u(t^{km}{\zeta_1}^k g)\right)$$

as

$$\exp\left(v(t^{kn}{\zeta_2}^k f)\right)\exp\left(u(t^{km}{\zeta_1}^k g)\right)\exp\left(-v(t^{kn}{\zeta_2}^k f)\right).$$

Then, using the same idea as above, we compute $\overline{D}\Phi_4$. Let

$$\sigma_1 = v(t^{kn}{\zeta_2}^k f) \tag{3.2.14}$$

$$\sigma_2 = u(t^{km}{\zeta_1}^k g) \tag{3.2.15}$$

$$\sigma_3 = -v(t^{kn}{\zeta_2}^k f). \tag{3.2.16}$$

Thus $\overline{D}\Phi_4 = (\overline{D}\sigma_1 + \exp \operatorname{ad} \sigma_1(\overline{D}\sigma_2 + \exp \operatorname{ad} \sigma_2(\overline{D}\sigma_3)))\Phi_4$. Straightforward computations show:

$$\begin{aligned}
\overline{D}\sigma_3 &= -v(t^{kn}{\zeta_2}^k f) - v(t^{kn}{\zeta_2}^k(\overline{D}f)), \\
\exp \operatorname{ad} \sigma_2 \cdot \overline{D}\sigma_3 &= -v(t^{kn}{\zeta_2}^k f) + [v,u](t^{km+kn}{\zeta_1}^k{\zeta_2}^k fg) \\
&\quad + \langle v,u\rangle kn\delta_{m+n,\,0}{\zeta_1}^k{\zeta_2}^k fgc + u(t^{2km+kn}{\zeta_1}^{2k}{\zeta_2}^k f(g)^2) \\
&\quad - v(t^{kn}{\zeta_2}^k(\overline{D}f)) + [v,u](t^{km+kn}{\zeta_1}^k{\zeta_2}^k(\overline{D}f)g) \\
&\quad + \langle v,u\rangle kn\delta_{m+n,\,0}{\zeta_1}^k{\zeta_2}^k(\overline{D}f)gc \\
&\quad + u(t^{2km+kn}{\zeta_1}^{2k}{\zeta_2}^k(\overline{D}f)(g)^2), \quad \text{and} \\
\overline{D}\sigma_2 &= u(t^{km}{\zeta_1}^k g) + u(t^{km}{\zeta_1}^k(\overline{D}g)).
\end{aligned}$$

Thus

$$\begin{aligned}
&\overline{D}\sigma_2 + \exp \operatorname{ad} \sigma_2 \cdot (\overline{D}\sigma_3) = u(t^{km}{\zeta_1}^k g + t^{2km+kn}{\zeta_1}^{2k}{\zeta_2}^k f(g)^2) \\
&\quad - v(t^{kn}{\zeta_2}^k f) + [v,u](t^{km+kn}{\zeta_1}^k{\zeta_2}^k fg) + \langle v,u\rangle kn\delta_{m+n,\,0}{\zeta_1}^k{\zeta_2}^k fgc \\
&\quad + u(t^{km}{\zeta_1}^k(\overline{D}g) + t^{2km+kn}{\zeta_1}^{2k}{\zeta_2}^k(\overline{D}f)(g)^2) - v(t^{kn}{\zeta_2}^k(\overline{D}f)) \\
&\quad + [v,u](t^{km+kn}{\zeta_1}^k{\zeta_2}^k(\overline{D}f)g) + \langle v,u\rangle kn\delta_{m+n,\,0}{\zeta_1}^k{\zeta_2}^k(\overline{D}f)gc.
\end{aligned}$$

Continuing we see:

$$\begin{aligned}
&\exp \operatorname{ad} \sigma_1(\overline{D}\sigma_2 + \exp \operatorname{ad} \sigma_2(\overline{D}\sigma_3)) = u(t^{km}{\zeta_1}^k g + t^{2km+kn}{\zeta_1}^{2k}{\zeta_2}^k f(g)^2) \\
&\quad + ([v,u])(2t^{km+kn}{\zeta_1}^k{\zeta_2}^k fg + t^{2km+2kn}{\zeta_1}^{2k}{\zeta_2}^{2k}(fg)^2) \\
&\quad + \langle v,u\rangle kn\delta_{m+n,\,0}(2{\zeta_1}^k{\zeta_2}^k fg + {\zeta_1}^{2k}{\zeta_2}^{2k}(fg)^2)c \\
&\quad - v(3t^{km+2kn}{\zeta_1}^k{\zeta_2}^{2k}(f)^2 g + t^{2km+3kn}{\zeta_1}^{2k}{\zeta_2}^{3k}(f)^3(g)^2 + t^{kn}{\zeta_2}^k f)
\end{aligned}$$

$$+u(t^{km}\zeta_1{}^k(\overline{D}g) + t^{2km+kn}\zeta_1{}^{2k}\zeta_2{}^k(\overline{D}f)(g)^2)$$
$$+[v,u](t^{km+kn}\zeta_1{}^k\zeta_2{}^k\overline{D}(fg) + t^{2km+2kn}\zeta_1{}^{2k}\zeta_2{}^{2k}f(\overline{D}f)(g)^2)$$
$$+\langle v,u\rangle nk\delta_{m+n,\,0}(\zeta_1{}^k\zeta_2{}^k\overline{D}(fg) + \zeta_1{}^{2k}\zeta_2{}^{2k}f(g)^2(\overline{D}f))c$$
$$-v\left(t^{km+2kn}\zeta_1{}^k\zeta_2{}^{2k}\overline{D}((f)^2g) + t^{2km+3kn}\zeta_1{}^{2k}\zeta_2{}^{3k}(fg)^2\overline{D}f\right.$$
$$\left.+t^{kn}\zeta_2{}^k(\overline{D}f)\right),$$

and

$$\overline{D}\sigma_1 = v(t^{kn}\zeta_2{}^kf) + v(t^{nk}\zeta_2{}^k(\overline{D}f)).$$

Thus we have

$$\begin{aligned}
\overline{D}\Phi_4 &= (\overline{D}\sigma_1 + \exp\operatorname{ad}\sigma_1(\overline{D}\sigma_2 + \exp\operatorname{ad}\sigma_2(\overline{D}\sigma_3)))\Phi_4 \qquad (3.2.17)\\
&= \Big(u(t^{km}\zeta_1{}^kg + t^{2km+kn}\zeta_1{}^{2k}\zeta_2{}^k(f(g)^2))\\
&\quad +[v,u](2t^{km+kn}\zeta_1{}^k\zeta_2{}^kfg + t^{2km+2kn}\zeta_1{}^{2k}\zeta_2{}^{2k}(fg)^2)\\
&\quad +\langle v,u\rangle kn\delta_{m+n,\,0}(2\zeta_1{}^k\zeta_2{}^kfg + \zeta_1{}^{2k}\zeta_2{}^{2k}(fg)^2)c\\
&\quad -v(3t^{km+2kn}\zeta_1{}^k\zeta_2{}^{2k}(f)^2g + t^{2km+3kn}\zeta_1{}^{2k}\zeta_2{}^{3k}(f)^3(g)^2)\\
&\quad +u(t^{km}\zeta_1{}^k(\overline{D}g) + t^{2km+kn}\zeta_1{}^{2k}\zeta_2{}^k(\overline{D}f)(g)^2)\\
&\quad +[v,u](t^{km+kn}\zeta_1{}^k\zeta_2{}^k\overline{D}(fg) + t^{2km+2kn}\zeta_1{}^{2k}\zeta_2{}^{2k}f(g)^2(\overline{D}f))\\
&\quad +\langle v,u\rangle kn\delta_{m+n,\,0}(\zeta_1{}^k\zeta_2{}^k\overline{D}(fg) + \zeta_1{}^{2k}\zeta_2{}^{2k}f(g)^2(\overline{D}f))c\\
&\quad -v(t^{km+2kn}\zeta_1{}^k\zeta_2{}^{2k}\overline{D}((f)^2g) + t^{2km+3kn}\zeta_1{}^{2k}\zeta_2{}^{3k}(fg)^2(\overline{D}f))\Big)\,\Phi_4\\
&= \tilde{\phi_4}\Phi_4.
\end{aligned}$$

So using (3.2.17) we replace (3.2.13) with

$$\overline{D}\Psi = \left(\overline{D}\phi_1 + \exp\operatorname{ad}\phi_1\left(\overline{D}\phi_2 + \exp\operatorname{ad}\phi_2\left(\overline{D}\phi_3 + \exp\operatorname{ad}\phi_3\left(\tilde{\phi_4}\right)\right)\right)\right)\Psi.$$

Further computations show:

$$\exp\operatorname{ad}\phi_3(\tilde{\phi_4}) = u((t^{km}\zeta_1{}^kg - t^{2km+kn}\zeta_1{}^{2k}\zeta_2{}^kf(g)^2)p^2)$$
$$+[v,u](2t^{km+kn}\zeta_1{}^k\zeta_2{}^kfgp) + \langle v,u\rangle kn\delta_{m+n,\,0}(2\zeta_1{}^k\zeta_2{}^kfgp)c \qquad (3.2.18)$$

$$+v(-3t^{km+2kn}\zeta_1{}^k\zeta_2{}^{2k}(f)^2g - t^{2km+3kn}\zeta_1{}^{2k}\zeta_2{}^{3k}(f)^3(g)^2)$$
$$+u\Big(t^{km}\zeta_1{}^k(\overline{D}g) + t^{2km+kn}\zeta_1{}^{2k}\zeta_2{}^k(g)^2(\overline{D}f) - 2t^{2km+kn}\zeta_1{}^{2k}\zeta_2{}^k g(\overline{D}fg)p$$
$$-2t^{3km+2kn}\zeta_1{}^{3k}\zeta_2{}^{2k}f(g)^3(\overline{D}f)p + t^{3km+2kn}\zeta_1{}^{3k}\zeta_2{}^{2k}(g)^2\overline{D}((f)^2g)p^2$$
$$+t^{4km+3kn}\zeta_1{}^{4k}\zeta_2{}^{3k}(f)^2(\overline{D}f)(g)^4p^2\Big)$$
$$+[v,u](t^{km+kn}\zeta_1{}^k\zeta_2{}^k\overline{D}(fg) + t^{2km+2kn}\zeta_1{}^{2k}\zeta_2{}^{2k}f(g)^2(\overline{D}f)$$
$$-t^{2km+2kn}\zeta_1{}^{2k}\zeta_2{}^{2k}g(\overline{D}((f)^2g))p - t^{3km+3kn}\zeta_1{}^{3k}\zeta_2{}^{3k}(f)^2(g)^3(\overline{D}f)p)$$
$$+\langle v,u\rangle kn\delta_{m+n,0}(\zeta_1{}^k\zeta_2{}^k\overline{D}(fg) + \zeta_1{}^{2k}\zeta_2{}^{2k}f(g)^2(\overline{D}f)$$
$$-\zeta_1{}^{3k}\zeta_2{}^{3k}(f)^2(g)^3(\overline{D}f)p)c$$
$$-v(t^{km+2kn}\overline{D}((f)^2g) + t^{2km+3kn}\zeta_1{}^{2k}\zeta_2{}^{3k}(fg)^2(\overline{D}f)),$$

and

$$\overline{D}\phi_3 = -u(t^{km}\zeta_1{}^k gp - 2t^{2km+kn}\zeta_1{}^{2k}\zeta_2{}^k f(g)^2p^2)$$
$$-u(t^{km}\zeta_1{}^k(\overline{D}g)p - t^{2km+kn}\zeta_1{}^{2k}\zeta_2{}^k g(\overline{D}(fg))p^2). \qquad (3.2.19)$$

Thus (3.2.18) and (3.2.19) give

$$\overline{D}\phi_3 + \exp \operatorname{ad} \phi_3(\tilde{\phi}_4) = ([v,u])(2t^{km+kn}\zeta_1{}^k\zeta_2{}^k fgp)$$
$$+\langle v,u\rangle kn\delta_{m+n,0}(2\zeta_1{}^k\zeta_2{}^k fgp)c \qquad (3.2.20)$$
$$+v(-3t^{km+2kn}\zeta_1{}^k\zeta_2{}^{2k}(f)^2g - t^{2km+3kn}\zeta_1{}^{2k}\zeta_2{}^{3k}(f)^3(g)^2)$$
$$+[v,u](t^{km+kn}\zeta_1{}^k\zeta_2{}^k\overline{D}(fg) - t^{2km+2kn}\zeta_1{}^{2k}\zeta_2{}^{2k}f(g)^2(\overline{D}f)p$$
$$-t^{2km+2kn}\zeta_1{}^{2k}\zeta_2{}^{2k}(f)^2g(\overline{D}g)p)$$
$$+\langle v,u\rangle kn\delta_{m+n,0}(\zeta_1{}^k\zeta_2{}^k\overline{D}(fg) - \zeta_1{}^{2k}\zeta_2{}^{2k}f(g)^2(\overline{D}f)p - \zeta_1{}^{2k}\zeta_2{}^{2k}(f)^2g(\overline{D}g)p)c$$
$$-v(t^{km+2kn}\zeta_1{}^k\zeta_2{}^{2k}\overline{D}((f)^2g) + t^{2km+3kn}\zeta_1{}^{2k}\zeta_2{}^{3k}(fg)^2(\overline{D}f)).$$

Next we have

$$\exp \operatorname{ad} \phi_2(\overline{D}\phi_3 + \exp \operatorname{ad} \phi_3(\tilde{\phi}_4)) = [v,u](2t^{km+kn}\zeta_1{}^k\zeta_2{}^k fgp)$$
$$+\langle v,u\rangle kn\delta_{m+n,0}(2\zeta_1{}^k\zeta_2{}^k fgp)c \qquad (3.2.21)$$
$$+v(-3t^{km+2kn}\zeta_1{}^k\zeta_2{}^{2k}(f)^2gp^2 - t^{2km+3kn}\zeta_1{}^{2k}\zeta_2{}^{3k}(f)^3(g)^2p^2)$$

$$+[v,u](t^{km+kn}\zeta_1{}^k\zeta_2{}^k\overline{D}(fg) - t^{2km+2kn}\zeta_1{}^{2k}\zeta_2{}^{2k}(f)^2g(\overline{D}g)p$$
$$-t^{2km+2kn}\zeta_1{}^{2k}\zeta_2{}^{2k}f(g)^2(\overline{D}f)p)$$
$$+\langle v,u\rangle kn\delta_{m+n,\,0}(\zeta_1{}^k\zeta_2{}^k\overline{D}(fg) - \zeta_1{}^{2k}\zeta_2{}^{2k}f(g)^2(\overline{D}f)p$$
$$-\zeta_1{}^{2k}\zeta_2{}^{2k}(f)^2g(\overline{D}g)p)c$$
$$-v(t^{km+2kn}\zeta_1{}^k\zeta_2{}^{2k}(\overline{D}((f)^2g))p^2 + t^{2mk+3nk}\zeta_1{}^{2k}\zeta_2{}^{3k}(fg)^2(\overline{D}f)p^2),$$

and

$$\overline{D}\phi_2 = [v,u](-2t^{km+kn}\zeta_1{}^k\zeta_2{}^kfgp) + \langle v,u\rangle kn\delta_{m+n,\,0}(-2\zeta_1{}^k\zeta_2{}^kfgp)c$$
$$+[v,u](-t^{km+kn}\zeta_1{}^k\zeta_2{}^k(\overline{D}(fg))p) \tag{3.2.22}$$
$$+\langle v,u\rangle kn\delta_{m+n,\,0}(-\zeta_1{}^k\zeta_2{}^k(\overline{D}(fg))p)c.$$

Thus (3.2.21) and (3.2.22) imply

$$\overline{D}\phi_2 + \exp\operatorname{ad}\phi_2(\overline{D}\phi_3 + \exp\operatorname{ad}\phi_3(\tilde{\phi}_4)) = \tag{3.2.23}$$
$$= v(-3t^{km+2kn}\zeta_1{}^k\zeta_2{}^{2k}(f)^2gp^2 - t^{2km+3kn}\zeta_1{}^{2k}\zeta_2{}^{3k}(f)^3(g)^2p^2)$$
$$-v(t^{km+2kn}\zeta_1{}^k\zeta_2{}^{2k}(\overline{D}((f)^2g))p^2 + t^{2km+3kn}\zeta_1{}^{2k}\zeta_2{}^{3k}(fg)^2(\overline{D}f)p^2).$$

We also have

$$\overline{D}\phi_1 = v(3t^{km+2kn}\zeta_1{}^k\zeta_2{}^{2k}(f)^2gp - 2t^{2km+3kn}\zeta_1{}^{2k}\zeta_2{}^{3k}(f)^3(g)^2p^2) \tag{3.2.24}$$
$$+v(t^{km+2kn}\zeta_1{}^k\zeta_2{}^{2k}(\overline{D}((f)^2g))p - t^{2km+3kn}\zeta_1{}^{2k}\zeta_2{}^{3k}(f)^2g(\overline{D}(fg))p^2).$$

Combining (3.2.18) through (3.2.24) with the fact that

$$3t^{km+2kn}\zeta_1{}^k\zeta_2{}^{2k}f^2gp - 2t^{2km+3kn}\zeta_1{}^{2k}\zeta_2{}^{3k}f^3g^2p^2$$
$$-3t^{km+2kn}\zeta_1{}^k\zeta_2{}^{2k}f^2gp^2 - t^{2km+3kn}\zeta_1{}^{2k}\zeta_2{}^{3k}f^3g^2p^2 = 0$$

and

$$-t^{km+2kn}\zeta_1{}^k\zeta_2{}^{2k}(\overline{D}((f)^2g))p^2 - t^{2km+3kn}\zeta_1{}^{2k}\zeta_2{}^{3k}(fg)^2(\overline{D}f)p^2$$
$$+t^{km+2kn}\zeta_1{}^k\zeta_2{}^{2k}(\overline{D}((f)^2g))p - t^{2km+3kn}\zeta_1{}^{2k}\zeta_2{}^{3k}(f)^2g(\overline{D}(fg))p^2 = 0,$$

we see $\overline{D}\Psi = 0$. Since Ψ is a product of power series, each with constant term one, we must have $\Psi = 1$.□

Corollary 3.2.3 *Let $u = x_a$ and $v = x_b$ with $a, b \in \hat{\Delta}$, and $\langle \overline{a}, \overline{b} \rangle = -2$. Then for $m, n \in \mathbf{Z}$ we have*

$$\begin{aligned} &\exp(v(t^n\zeta_2))\exp(u(t^m\zeta_1)) = \\ &\quad \exp\left(u(t^m\zeta_1(1+t^{m+n}\zeta_1\zeta_2)^{-1})\right)\exp\left([v,u](\log(1+t^{m+n}\zeta_1\zeta_2))\right. \\ &\quad\quad \left.+\langle v,u\rangle n\delta_{m+n,\,0}\log(1+\zeta_1\zeta_2)c\right)\exp(v(t^n\zeta_2(1+t^{m+n}\zeta_1\zeta_2)^{-1})). \end{aligned}$$

Proposition 3.2.4 *Let $u = x_a$ and $v = \alpha_i + \delta_{i,0}c$ with $\langle \overline{a}, \overline{a} \rangle = 2$, and $\alpha_i \in \Pi$. Then for $m, n \in \mathbf{Z}$ we have*

$$\begin{aligned} &\exp\left(-v(\log(1-t^{kn}{\zeta_2}^k))\right)\exp\left(u(t^{km}{\zeta_1}^k)\right) = \\ &\quad \exp\left(u(t^{km}{\zeta_1}^k(1-t^{kn}{\zeta_2}^k)^{-\langle\alpha_i,\overline{a}\rangle})\right)\exp\left(-v(\log(1-t^{kn}{\zeta_2}^k))\right). \end{aligned} \tag{3.2.25}$$

Proof: In this case,

$$\exp(\mathrm{ad}\, -v(\log(1-t^{kn}{\zeta_2}^k))\cdot u(t^{km}{\zeta_1}^k)) = \sum_{l\geq 0} -\frac{\langle\alpha_i,\overline{a}\rangle^l(\log(1-t^{kn}{\zeta_2}^k))^l}{l!}\cdot u(t^{km+ln}{\zeta_1}^k{\zeta_2}^l).$$

Thus by Theorem 3.1.5, the left hand side of (3.2.25) is given by

$$\exp\left(u\left(t^{km}{\zeta_1}^k\left(\sum_{l\geq 0} -\frac{\langle\alpha_i,\overline{a}\rangle^l(\log(1-t^{kn}{\zeta_2}^k))^l}{l!}\right)\right)\right)\exp\left(-v(\log(1-t^{kn}{\zeta_2}^k))\right).\square$$

Proposition 3.2.5 *Let $u = \alpha_i$ and $v = \alpha_j$ where $\alpha_i, \alpha_j \in \Pi$. Then*

$$\begin{aligned} &\exp\left(-v(\log(1-t^{-kn}{\zeta_2}^k))\right)\exp\left(-u(\log(1-t^{km}{\zeta_1}^k))\right) = \\ &\quad \exp\left(-u(\log(1-t^{km}{\zeta_1}^k))\right)\exp\left(\mu\langle v,u\rangle\log(1-{\zeta_1}^{kn'}{\zeta_2}^{km'})c\right)\times \\ &\quad \exp\left(-v(\log(1-t^{-nk}{\zeta_2}^k))\right), \end{aligned} \tag{3.2.26}$$

where $m, n \in \mathbf{Z}$, $m, n \geq 0$, $\mu = \gcd(km, kn)$, $m' = \frac{m}{\mu}$ and $n' = \frac{n}{\mu}$.

Proof: Here we have $[v,u] = 0$. Thus Theorem 3.1.5 gives the left hand side of (3.2.26) as

$$\begin{aligned} &\exp\left(-u(\log(1-t^{km}{\zeta_1}^k))\right)\times \\ &\quad\times\exp\left(\sum_{j,\,l\geq 1}\frac{\langle v,u\rangle(-nj)}{nl}\delta_{-nj+ml,0}c{\zeta_1}^{kl}{\zeta_2}^{kj}\right)\exp\left(-v(\log(1-t^{-kn}{\zeta_2}^k))\right). \end{aligned}$$

The result follows (cf. [M]). □

Lemma 3.2.6 *Let $u \in \underline{g}$, and let $f(\zeta)$, $g(\zeta) \in \zeta\mathbf{C}[[\zeta]]$. Then for k, $l \geq 0$ and k, l, $m \in \mathbf{Z}$ we have*

$$\left(ad\,(-d)\cdot f({\zeta_2}^k)\right)^l \cdot v(g(t^{km}{\zeta_1}^k)) = \sum_{j\geq 1}(-1)^l a_j(kmj)^l v\otimes t^{kmj}{\zeta_1}^{kj}(f({\zeta_2}^k))^l$$

where $g({\zeta_1}^k) = \sum_{j\geq 1} a_j {\zeta_1}^{kj}$.

Proof: Consider the case $l = 1$. By (2.1.11), (2.1.12) and (3.1.4), we see that

$$\begin{aligned}\text{ad}\,((-d)\cdot f({\zeta_2}^k))\cdot v(g(t^{km}{\zeta_1}^k)) &= [-d, \textstyle\sum_{j\geq 1} a_j v\otimes t^{kmj}{\zeta_1}^{kj}]f({\zeta_2}^k) \\ &= \left(\sum_{j\geq 1}(-1)^l(kmj)a_j v\otimes t^{kmj}{\zeta_1}^{kj}\right) f({\zeta_2}^k).\end{aligned}$$

Now apply induction on l. □

Proposition 3.2.7 *Let $u \in \underline{g}$ with $f(\zeta) \in \zeta\mathbf{C}[[\zeta]]$. Then*

$$\begin{aligned}\exp\left(-d\cdot(\log(1-{\zeta_2}^k))\right)\exp\left(u(f(t^{km}{\zeta_1}^k))\right) &= = \exp\left(u(f(t^{km}{\zeta_1}^k(1-{\zeta_2}^k)^{-m}))\right)\times \\ &\quad \times\exp\left(-d\cdot(\log(1-{\zeta_2}^k))\right).\end{aligned}$$

Proof: By Lemma 3.1.2 and Lemma 3.2.6, we have

$$\begin{aligned}&\exp\left(\exp\left(\text{ad}\,(-d)\right))\log(1-{\zeta_2}^k)\right)\cdot u(f(t^{km}{\zeta_1}^k)) \\ &= \exp\left(\sum_{l\geq 0}\frac{\left(\text{ad}\,(-d)\cdot(\log(1-{\zeta_2}^k))\right)^l}{l!}\cdot\sum_{j\geq 1} a_j u\otimes t^{kmj}{\zeta_1}^{kj}\right) \\ &= \exp\left(\sum_{\substack{l\geq 0\\ j\geq 1}}\frac{(-1)^l(kmj)^l a_j u\otimes t^{kmj}{\zeta_1}^{kj}(\log(1-{\zeta_2}^k))^l}{l!}\right) \\ &= \exp\left(\sum_{j\geq 1}\left(\sum_{l\geq 0}\frac{(-1)^l(kmj)^l(\log(1-{\zeta_2}^k))^l a_j u\otimes t^{kmj}{\zeta_1}^{kj}}{l!}\right)\right) \\ &= \exp\left(u(f(t^{km}{\zeta_1}^k(1-{\zeta_2}^k)^{-m}))\right).\end{aligned}$$

Now apply Lemma 3.1.1. □

In particular, Proposition 3.2.7 implies that for $a \in \hat{\Delta}$

$$\exp\left(-d\log(1-\zeta_2)\right)\exp\left(x_a(t^m\zeta_1)\right) = \exp\left(x_a(t^m\zeta_1(1-\zeta_2)^{-m})\right)\exp\left(-d(\log(1-\zeta_2))\right), \tag{3.2.27}$$

and for $(n, i) \in \hat{I}$, that

$$\exp(-d\log(1-\zeta_2))\exp(-\alpha_i(\log(1-t^n\zeta_1)) - \delta_{i,0}c\log(1-\zeta_1))$$
$$= \exp(-\alpha_i(\log(1-t^n\zeta_1(1-\zeta_2)^{-n})) - \delta_{i,0}c\log(1-\zeta_1))\exp(-d\log(1-\zeta_2)). \tag{3.2.28}$$

The preceding propositions allow us to "straighten" all of the cases of C_1 and C_2 for the type 1 algebras of type A, D or E; thus we have proved Theorem 3.1.6 in this setting.

3.3 Exponential identities for $B_n^{(1)}$, $C_n^{(1)}$ and $F_4^{(1)}$

In this section we will develop identities analogous to those in §3.2 for the type one affine algebras $B_n^{(1)}$, $C_n^{(1)}$ or $F_4^{(1)}$. The identities for these remaining type one affines are computed directly using the identities for the simply-laced affines and the Baker-Campbell-Hausdorff formula. At this time the reader may wish to recall the setting of §2.2 and all its notation.

We state the exponential identities as corollaries to those found in the preceding section.

Corollary 3.3.1 *Let $u = x_a$ and $v = x_b$ with $\overline{a}, \overline{b} \in \Delta$. If $\overline{a}, \overline{b} \in L$ are such that*

$$\langle \overline{a}, \overline{b}\rangle = -1,\ \langle \overline{a}_{[0]}, \overline{a}_{[0]}\rangle \leq \langle \overline{b}_{[0]}, \overline{b}_{[0]}\rangle \text{ and } \langle \overline{a}, \overline{b}\rangle \leq \langle \nu\overline{a}, \overline{b}\rangle,$$

then for $m, n \in \mathbf{Z}$, we have

$$\begin{aligned}&\exp\left(v_{[0]}^{+}(t^n\zeta_2)\right)\exp\left(u_{[0]}^{+}(t^m\zeta_1)\right) = \\ &\exp\left(u_{[0]}^{+}(t^m\zeta_1)\right)\exp\left(-([u_{[0]}^{+}, v_{[0]}^{+}](t^{m+n}\zeta_1\zeta_2)\right) \times \\ &\exp\left(\tfrac{1}{2}([u_{[0]}^{+}, [u_{[0]}^{+}, v_{[0]}^{+}]](t^{2m+n}{\zeta_1}^2\zeta_2)\right)\exp\left(v_{[0]}^{+}(t^n\zeta_2)\right).\end{aligned} \tag{3.3.1}$$

Proof: A cases by case argument exploiting Remark 2.2.1, (3.1.2), Theorem 2.2.2 and Theorem 3.2.1 gives the result. □

Corollary 3.3.2 *Let $u = x_a$ and $v = x_b$ with $\overline{a}$, $\overline{b} \in \Delta$ be such that*

$$\langle \overline{a}, \overline{a} \rangle = \langle \overline{b}, \overline{b} \rangle = -\langle \overline{a}, \overline{b} \rangle = 2.$$

Then for m, $n \in \mathbf{Z}$, we have

$$\begin{aligned} &\exp\left(v^+_{[0]}(t^n\zeta_2)\right)\exp\left(u^+_{[0]}(t^m\zeta_1)\right) = \\ &\quad \exp\left(u^+_{[0]}(t^m\zeta_1(1+t^{m+n}\zeta_1\zeta_2)^{-1})\right)\exp\left(([v^+_{[0]}, u^+_{[0]}])(\log(1+t^{m+n}\zeta_1\zeta_2)) + \right. \\ &\quad \left. \langle v^+_{[0]}, u^+_{[0]}\rangle n\delta_{m+n,\,0}\log(1+\zeta_1\zeta_2)c\right)\exp\left(v^+_{[0]}(t^n\zeta_2(1+t^{m+n}\zeta_1\zeta_2)^{-1})\right). \end{aligned} \tag{3.3.2}$$

Proof: One needs to consider the cases $\hat{\nu}a = a$ and $\hat{\nu}a \neq a$. The first case is just Proposition 3.2.2. In the second case, Remark 2.2.1, (3.1.2) and Proposition 3.2.2 yield the result. □

Corollary 3.3.3 *Let $u = x_a$ and $v = \alpha_i + \delta_{i,0}c$ with $\langle \overline{a}, \overline{a} \rangle = 2$ and $\alpha_i \in \Pi \cup \{\alpha_0\}$. Then for m, $n \in \mathbf{Z}$, we have*

$$\begin{aligned} &\exp\left(-v^+_{[0]}(\log(1-t^n\zeta_2))\right)\exp\left(u^+_{[0]}(t^m\zeta_1)\right) \\ &\quad = \exp\left(u^+_{[0]}(t^m\zeta_1(1-t^n\zeta_2)^{-\langle \alpha_{[0]}, \alpha^+_{i,\,[0]}\rangle})\right) \times \\ &\quad \exp\left(-v^+_{[0]}(\log(1-t^n\zeta_2))\right). \end{aligned} \tag{3.3.3}$$

Proof: Apply Corollary 3.2.4, and (3.1.2). □

Corollary 3.3.4 *Let $u = \alpha_i \otimes 1$ and $v = \alpha_j \otimes 1$ with α_i, $\alpha_j \in \Pi$. Then we have*

$$\begin{aligned} &\exp\left(-v^+_{[0]}(\log(1-t^{-n}\zeta_2))\right)\exp\left(-u^+_{[0]}(\log(1-t^m\zeta_1))\right) \\ &\quad = \exp\left(-u^+_{[0]}(\log(1-t^m\zeta_1))\right)\exp\left(\mu\langle v^+_{[0]}, u^+_{[0]}\rangle\log(1-{\zeta_1}^{n'}{\zeta_2}^{m'})c\right) \times \\ &\quad \exp\left(-v^+_{[0]}(\log(1-t^{-n}\zeta_2))\right), \end{aligned}$$

where m, $n \in \mathbf{Z}$, $\mu = \gcd(m,n)$, $m' = \frac{m}{\mu}$ and $n' = \frac{n}{\mu}$.

Proof: Apply Proposition 3.2.4 and (3.1.2). □

Corollary 3.3.5 *Let u be an element of $\underline{\mathfrak{g}}$ and let $f(\zeta) \in \zeta\mathbf{C}[[\zeta]]$. Then for $m \in \mathbf{Z}$, we have*

$$\exp\left(-d(\log(1-\zeta_2))\right)\exp\left(u^+_{[0]}(f(t^m\zeta_1))\right) =$$
$$\exp\left(u^+_{[0]}(f(t^m\zeta_1(1-\zeta_2)^{-m}))\right)\exp\left(-d(\log(1-\zeta_2))\right).$$

Proof: Apply Proposition 3.2.7 and (3.1.2). □

We now have proved Theorem 3.1.6 for $B_n^{(1)}$, $C_n^{(1)}$ and $F_4^{(1)}$.

3.4 Exponential identities for $G_2^{(1)}$

We now consider the case where Δ and the automorphism ν are as in §2.4. We also retain the notation from the same section. Again the identities are derived using the machinery set up in Sections 2.4, 3.1 and 3.2. We note that (3.4.1) corrects a misprint in Lemma 4.3.4 (i) of [M].

Corollary 3.4.1 *Let $u = x_a$ and $v = x_b$ be elements of $\underline{\mathfrak{g}}$ such that*

$$\langle \bar{a}, \bar{b}\rangle = -1,\ \langle \bar{a}_{[0]}, \bar{a}_{[0]}\rangle \le \langle \bar{b}_{[0]}, \bar{b}_{[0]}\rangle \text{ and } \langle \bar{a}, \bar{b}\rangle \le \langle \nu\bar{a}, \bar{b}\rangle \le \langle \nu^2\bar{a}, \bar{b}\rangle.$$

Then for m, $n \in \mathbf{Z}$, we have

$$\begin{aligned}
&\exp\left(v^+_{[0]}(t^n\zeta_2)\right)\exp\left(u^+_{[0]}(t^m\zeta_1)\right) = \\
&= \exp\left(u^+_{[0]}(t^m\zeta_1)\right)\exp\left(-([u^+_{[0]}, v^+_{[0]}])(t^{m+n}\zeta_1\zeta_2)\right) \times \\
&\quad \exp\left(\tfrac{1}{2}([u^+_{[0]}, [u^+_{[0]}, v^+_{[0]}]](t^{2m+n}{\zeta_1}^2\zeta_2)\right) \times \\
&\quad \exp\left(-\tfrac{1}{2}([v^+_{[0]}, [u^+_{[0]}, v^+_{[0]}]](t^{m+2n}\zeta_1{\zeta_2}^2)\right) \times \\
&\quad \exp\left(-\tfrac{1}{3!}([u^+_{[0]}, [u^+_{[0]}, [u^+_{[0]}, v^+_{[0]}]]](t^{3m+n}{\zeta_1}^3\zeta_2)\right) \times \\
&\quad \exp\left(-\tfrac{1}{3}([v^+_{[0]}, [u^+_{[0]}, [u^+_{[0]}, [u^+_{[0]}, v^+_{[0]}]]]](t^{3m+2n}{\zeta_1}^3{\zeta_2}^2)\right) \times \\
&\quad \exp\left(v^+_{[0]}(t^n\zeta_2)\right).
\end{aligned} \tag{3.4.1}$$

Proof: A case by case argument using (3.1.2), Proposition 2.4.1, Theorem 2.4.2 and Proposition 3.2.1 yields (3.4.1) (cf. (2.4.7)). □

Corollary 3.4.2 *Let $u = x_a$ and $v = x_b$ be elements of $\underline{g}$ such that $\langle \overline{a}, \overline{b} \rangle = -2$. Then for m, $n \in \mathbf{Z}$, we have*

$$\exp\left(v^+_{[0]}(t^n\zeta_2)\right)\exp\left(u^+_{[0]}(t^m\zeta_1)\right) =$$
$$= \exp\left(u^+_{[0]}(t^m\zeta_1\zeta_1(1+t^{m+n}\zeta_1\zeta_2)^{-1})\right)\exp\left(([v^+_{[0]}, u^+_{[0]}])(\log(1+t^{m+n}\zeta_1\zeta_2))+\right.$$
$$\left.\langle v^+_{[0]}, u^+_{[0]}\rangle n\delta_{m+n,\,0}\log(1+\zeta_1\zeta_2)c\right)\exp\left(v^+_{[0]}(t^n\zeta_2(1+t^{m+n}\zeta_1\zeta_2)^{-1})\right).$$

Proof: When $\overline{a} \in \Delta^0$, the statement is given by Corollary 3.2.3. For $\overline{a} \in \Delta^1$, use (3.1.2) and Proposition 3.2.3 to obtain the result. □

Corollary 3.4.3 *Let $u = x_a$ and $v = \alpha_i + \delta_{i,0}c$ be elements of $\underline{g}$ with $\alpha_i \in \Pi \cup \{\alpha_0\}$. Then for m, $n \in \mathbf{Z}$, we have*

$$\exp\left(-v^+_{[0]}(\log(1-t^n\zeta_2))\right)\exp\left(u^+_{[0]}(t^m\zeta_1\zeta_1)\right) =$$
$$\exp\left(u^+_{[0]}(t^m\zeta_1\zeta_1(1-t^n\zeta_2)^{-\langle \alpha_{[0]}, \alpha^+_{i,[0]}\rangle})\right)\exp\left(-v^+_{[0]}(\log(1-t^n\zeta_2))\right).$$

Proof: Apply Proposition 3.2.4 and (3.1.2). □

Corollary 3.4.4 *Let $u = \alpha_i$ and $v = \alpha_j$ with α_i, $\alpha_j \in \Pi$. Then we have*

$$\exp\left(-v^+_{[0]}(\log(1-t^n\zeta_2))\right)\exp\left(-u^+_{[0]}(\log(1-t^m\zeta_1))\right) =$$
$$= \exp\left(-u^+_{[0]}(\log(1-t^m\zeta_1))\right)\exp\left(\mu\langle v^+_{[0]}, u^+_{[0]}\rangle\log(1-\zeta_1{}^{n'}\zeta_2{}^{m'})c\right) \times$$
$$\exp\left(-v^+_{[0]}(\log(1-t^{-n}\zeta_2))\right),$$

where m, $n \in \mathbf{Z}_+$, $\mu = \gcd(m,n)$, $m' = \frac{m}{\mu}$, and $n' = \frac{n}{\mu}$.

Proof: Apply Proposition 3.2.4 and (3.1.2). □

Corollary 3.4.5 *Let $u \in \underline{g}$ and $f(\zeta) \in \zeta\mathbf{C}[[\zeta]]$. Then for $m \in \mathbf{Z}$, we have*

$$\exp\left(-d\cdot\log(1-\zeta_2)\right)\exp\left(u^+_{[0]}(f(t^m\zeta_1\zeta_1))\right) =$$
$$\exp\left(u^+_{[0]}(f(t^m\zeta_1(1-\zeta_2)^{-m}))\right)\exp\left(-d\cdot(\log(1-\zeta_2))\right).$$

Proof: Apply Proposition 3.2.7 and (3.1.2). □

Theorem 3.1.6 now follows for $G_2^{(1)}$.

3.5 Exponential identities for $A_{2n-1}^{(2)}$, $D_{n+1}^{(2)}$ and $E_6^{(2)}$

The reader may wish to review all the notation found in §2.3. Again the identities for $A_{2n-1}^{(1)}$, $D_{n+1}^{(1)}$ and $E_6^{(1)}$, respectively, along with the B-C-H formula allow us to directly compute the exponential identities needed for the algebras $A_{2n-1}^{(2)}$, $D_{n+1}^{(2)}$ and $E_6^{(2)}$, respectively.

We define

$$u^+(f(t^n\zeta)) = \sum_{j\geq 1} a_j u^+_{[mj]} \otimes t^{mj}\zeta^j, \tag{3.5.1}$$

where $u \in \underline{g}$, $f(\zeta) = \sum_{j\geq 1} a_j\zeta^j \in \zeta\mathbf{C}[[\zeta]]$ and $m \in \mathbf{Z}$ (cf. (3.1.6)). Note that for $u \in \underline{g}$ we have

$$u^+(f(t^m\zeta)) = \begin{cases} u(f(t^m\zeta)) & \text{if } \hat{\nu}u = u \text{ and } m \in 2\mathbf{Z} \\ u(f(t^m\zeta)) + \hat{\nu}u(f((-t)^m\zeta)) & \text{if } \hat{\nu}u \neq u. \end{cases} \tag{3.5.2}$$

Note that we do not need to consider $u \in \underline{g}$ such that $\hat{\nu}u = u$ and $m \in 2\mathbf{Z}+1$ (cf. Theorem 2.3.3 and Theorem 3.1.6).

Corollary 3.5.1 *Let $u = x_a$ and $v = x_b$ be elements of $\underline{g}$ such that $\langle \bar{a}, \bar{b}\rangle = -2$ and $\bar{a} \in \Delta^+$. Then for m, $n \in \mathbf{Z}$, we have*

$$\begin{aligned} &\exp(v^+(t^n\zeta_2))\exp(u^+(t^m\zeta_1)) = \\ &= \exp(u^+(t^m\zeta_1(1+t^{m+n}\zeta_1\zeta_2)^{-1}))\exp(-([u,v])^+(\log(1+t^{m+n}\zeta_1\zeta_2))) \times \\ &\quad \exp\left(\langle v^+_{[n]}, u^+_{[m]}\rangle n\delta_{m+n,\,0}\log(1+\zeta_1\zeta_2)c\right) \times \\ &\quad \exp(v^+(t^n\zeta_2(1+t^{m+n}\zeta_1\zeta_2)^{-1})). \end{aligned}$$

Proof: If we have $[m] = [n] = [0]$ and $\bar{a} \in \Delta^0$, then the result is given by Corollary 3.2.3. The other possibility, $\bar{a} \notin \Delta^0$, follows from (3.5.2), Proposition 3.2.2 and (3.1.2). □

Corollary 3.5.2 *Let $u = x_a$ and $v = \alpha$ be elements of $\underline{g}$ such that α, $\bar{a} \in \Delta$. Then for m, $n \in \mathbf{Z}$ such that $[n] = [0]$, we have*

$$\exp(-v^+(\log(1-t^n\zeta_2)))\exp(u^+(t^m\zeta_1)) =$$

$$= \exp\left(u^+(t^m\zeta_1(1-t^n\zeta_2)^{\langle\alpha^+_{[0]},\,\bar{a}_{[0]}\rangle})\right)\exp\left(-v^+(\log(1-t^n\zeta_2))\right).$$

Proof: A case by case proof using Proposition 3.2.4, (3.5.2) and (3.1.2) yields the result. □

Corollary 3.5.3 *Let $u = x_a$ and $v = \alpha$ be elements of $\underline{g}$ such that $\alpha,\ \bar{a} \in \Delta$. Then for $m,\ n \in \mathbf{Z}$ such that $[n] \neq [0]$, we have*

$$\exp\left(-v^+(\log(1-t^n\zeta_2))\right)\exp\left(u^+(t^m\zeta_1)\right) =$$
$$= \exp\left(u^+(t^m\zeta_1(1-t^n\zeta_2)^{-\langle\alpha,\bar{a}-\nu\bar{a}\rangle}(1-t^{2n}\zeta_2{}^2)^{-\langle\alpha,\nu\bar{a}\rangle})\right)\times$$
$$\exp\left(-v^+(\log(1-t^n\zeta_2))\right).$$

Proof: Again the result follows from Proposition 3.2.4, (3.5.2) and (3.1.2). □

Corollary 3.5.4 *Let $u = \alpha$ and $v = \beta$ with $\alpha,\ \beta \in \Delta$. Then we have*

$$\exp\left(-v^+(\log(1-t^{-n}\zeta_2))\right)\exp\left(-u^+(\log(1-t^m\zeta_1))\right) =$$
$$= \exp\left(-u^+(\log(1-t^m\zeta_1))\right)\exp\left(-v^+(\log(1-t^{-n}\zeta_2))\right)\times$$
$$\exp\left(\mu\langle\alpha^+_{[0]},\beta^+_{[0]}\rangle\log(1-\zeta_1{}^{n'}\zeta_2{}^{m'})c\right),$$

where $m,\ n \in \mathbf{Z}_+$, $[n] = 0$ or $[m] = 0$, $\mu = \gcd(m,n)$, $m = \mu m'$ and $n = \mu n'$.

Proof: Use Proposition 3.2.5, (3.5.2) and (3.1.2). □

Corollary 3.5.5 *Let $u = \alpha$ and $v = \beta$ be elements of $\underline{g}$ such that $\alpha,\ \beta \in \Delta$. Then we have*

$$\exp\left(-v^+(\log(1-t^{-n}\zeta_2))\right)\exp\left(-u^+(\log(1-t^m\zeta_1))\right) =$$
$$= \exp\left(-u^+(\log(1-t^m\zeta_1))\right)\exp\left(-v^+(\log(1-t^{-n}\zeta_2))\right)\times$$
$$\exp\left(2\mu\log(1-\zeta_1{}^{n'}\zeta_2{}^{m'})^{\langle\alpha,\beta-\nu\beta\rangle}(1-\zeta_1{}^{2n'}\zeta_2{}^{2m'})^{\langle\alpha,\nu\beta\rangle}c\right),$$

where $m, n \in \mathbf{Z}$ such that $[mn] \neq [0]$ and μ, m, n, m', and n' are as in Corollary 3.5.4.

Proof: Let $\alpha,\ \beta \in \Delta^1$. Then by (3.5.2), (3.1.2) and Proposition 3.2.5 we have

$$\begin{aligned}
&\exp\left(-v^+(\log(1-t^{-n}\zeta_2))\right)\exp\left(-u^+(\log(1-t^{m}\zeta_1))\right) = \\
&\quad \exp\left(-u^+(\log(1-t^{m}\zeta_1))\right)\exp\left(-v^+(\log(1-t^{-n}\zeta_2))\right) \times \\
&\quad \exp\Big(\mu(\log(1-\zeta_1{}^{n'}\zeta_2{}^{m'})^{\langle\alpha,\beta\rangle} + \mu\log(1-\zeta_1{}^{n'}(-\zeta_2)^{m'})^{\langle\alpha,\nu\beta\rangle} \\
&\qquad +\mu\log(1-(-\zeta_1)^{n'}\zeta_2{}^{m'})^{\langle\nu\alpha,\beta\rangle} + \mu\log(1-(-\zeta_1)^{n'}(-\zeta_2)^{m'})^{\langle\nu\alpha,\nu\beta\rangle})c\Big).
\end{aligned}$$

However the term in the last exponential may be replaced by

$$\begin{aligned}
&\mu\log\Big((1-\zeta_1{}^{n'}\zeta_2{}^{m'})^{2\langle\alpha,\beta-\nu\beta\rangle}(1-(-1)^{m'}\zeta_1{}^{n'}\zeta_2{}^{m'})^{\langle\alpha,\nu\beta\rangle}\times \\
&\quad \times(1-\zeta_1{}^{m'}\zeta_2{}^{n'})^{\langle\alpha,\nu\beta\rangle}(1-(-1)^{n'}\zeta_1{}^{n'}\zeta_2{}^{m'})^{\langle\alpha,\nu\beta\rangle}(1-\zeta_1{}^{n'}\zeta_2{}^{m'})^{\langle\alpha,\nu\beta\rangle}\Big)\,c.
\end{aligned}$$

This expression is further simplified to

$$2\mu\log\Big((1-\zeta_1{}^{n'}\zeta_2{}^{m'})^{\langle\alpha,\beta-\nu\beta\rangle}(1-\zeta_1{}^{2n'}\zeta_2{}^{2m'})^{\langle\alpha,\nu\beta\rangle}\Big)\,c.$$

Thus the result follows. □

Hence Theorem 3.1.6 is proved for $A^{(2)}_{2n-1}$, $D^{(2)}_{n+1}$ and $E^{(2)}_6$.

3.6 Exponential identities for $D_4^{(3)}$

In this section we will use the notations and definitions found in §3.4. Just as in the previous sections, the identities for $D_4^{(3)}$ are derived using the identities for $D_4^{(1)}$ and the B-C-H formula.

Define $u^+(f(t^n\zeta))$ to be given by (3.5.1). We note that

$$u^+(f(t^n\zeta)) = \begin{cases} u(f(t^n\zeta)) & \text{if } \hat{\nu}u = u \text{ and } n = 0(\bmod 3) \\ u(f(t^n\zeta)) + \hat{\nu}u(f((\omega^{-1}t)^n\zeta)) & \text{if } \hat{\nu}u \neq u. \\ \quad +\hat{\nu}^2u(f((\omega t)^n\zeta)) & \end{cases} \tag{3.6.1}$$

Corollary 3.6.1 *Let $u = x_a$ and $v = x_b$ be elements of $\underline{g}$ such that $\bar{a},\ \bar{b} \in \check{\Delta}$ and $\langle\bar{a},\bar{b}\rangle = 2$. Then for $m,\ n \in \mathbf{Z}$, we have*

$$\exp\left(-v^+(t^n\zeta_2)\right)\exp\left(u^+(t^m\zeta_1)\right) =$$

$$= \exp\left(u^+(t^m\zeta_1(1+t^{m+n}\zeta_1\zeta_2)^{-1})\right)\exp\left(-([u,v])^+(\log(1+t^{m+n}\zeta_1\zeta_2)^{-1})\right) \times$$
$$\exp\left(\langle v^+_{[n]}, u^+_{[m]}\rangle n\delta_{m+n,\,0}\log(1+\zeta_1\zeta_2)c\right)\exp\left(-v^+(t^n\zeta_2(1+t^{m+n}\zeta_1\zeta_2)^{-1})\right).$$

Proof: A case by case argument using Proposition 3.2.1, (3.6.1), (3.1.2), and yields the result. □

Corollary 3.6.2 *Let $u = x_a$ and $v = \alpha$ be elements of $\underline{\mathbf{g}}$ such that $\alpha,\ \bar{a} \in \check{\Delta}$. Then for $m,\ n \in \mathbf{Z}$ such that $[n] = [0]$, we have*

$$\exp\left(v^+(\log(1-t^n\zeta_2))\right)\exp\left(u^+(t^m\zeta_1)\right) =$$
$$\exp\left(u^+(t^m\zeta_1(1-t^n\zeta_2)^{-\langle\alpha^+_{[0]},\,\bar{a}_{[0]}\rangle})\right)\exp\left(v^+(\log(1-t^n\zeta_2))\right).$$

Proof: Use Proposition 3.2.4, (3.6.1) and (3.1.2) (cf. Corollary 3.5.2). □

Corollary 3.6.3 *Let $u = x_a$ and $v = \alpha$ be elements of $\underline{\mathbf{g}}$ such that $\alpha,\ \bar{a} \in \check{\Delta}$. Then for $m,\ n \in \mathbf{Z}$ such that $[n] \neq [0]$, we have*

$$\exp\left(-v^+(\log(1-t^n\zeta_2))\right)\exp\left(u^+(t^m\zeta_1)\right) =$$
$$\exp\left(u^+(t^m\zeta_1(1-t^n\zeta_2)^{\langle\alpha,\,\bar{a}-\nu\bar{a}\rangle}(1-t^{3n}\zeta_2{}^3)^{-\langle\alpha,\,\nu\bar{a}\rangle})\right) \times$$
$$\exp\left(-v^+(\log(1-t^n\zeta_2))\right).$$

Proof: See proof of Corollary 3.6.2. □

Corollary 3.6.4 *Let $u = \alpha$ and $v = \beta$ be elements of $\underline{\mathbf{g}}$ such that $\alpha,\ \beta \in \check{\Delta}$. Then we have*

$$\exp\left(-v^+(\log(1-t^n\zeta_2))\right)\exp\left(-u^+(\log(1-t^m\zeta_1))\right) =$$
$$= \exp\left(-u^+(\log(1-t^m\zeta_1))\right)\exp\left(-v^+(\log(1-t^n\zeta_2))\right) \times$$
$$\exp\left(\mu\langle\alpha^+_{[0]},\beta^+_{[0]}\rangle\log(1-\zeta_1{}^{n'}\zeta_2{}^{m'})c\right),$$

where $m,\ n \in \mathbf{Z}_+$ with $[m]$ or $[n] = 0$, $\mu = \gcd(m,n)$, $m = \mu m'$ and $n = \mu n'$.

Proof: Use Proposition 3.2.5., (3.6.1) and (3.1.2). □

Corollary 3.6.5 *Let u and v, m, n and μ be as in Corollary 3.6.4 with the exception that $[mn] \neq [0]$. Then we have*

$$\begin{aligned}&\exp\left(-v^+(\log(1-t^{-n}\zeta_2))\right)\exp\left(-u^+(\log(1-t^m\zeta_1))\right) = \\ &\quad = \exp\left(-u^+(\log(1-t^m\zeta_1))\right)\exp\left(-v^+(\log(1-t^{-n}\zeta_2))\right) \times \\ &\quad \exp\left(3\mu\log(1-{\zeta_1}^{n'}{\zeta_2}^{m'})^{\langle\alpha,\beta-\nu\beta\rangle}(1-{\zeta_1}^{3n'}{\zeta_2}^{3m'})^{\langle\alpha,\nu\beta\rangle}c\right).\end{aligned}$$

Proof: Apply (3.6.1), (3.1.2) and Proposition 3.2.4.□

We now have Theorem 3.1.6 for $D_4^{(3)}$.

3.7 Exponential identities for $A_{2n}^{(2)}$

We retain all notations and definitions from §2.6. Once again our tools for deriving the needed exponential identities are the formulas for the simply-laced algebra $A_{2n}^{(1)}$ and the B-C-H Theorem. We note that Corollary 3.7.2 corrects a misprint found in Lemma 4.4.1 (ii) in [M].

Before we begin to derive the exponential identities associated with $A_{2n}^{(2)}$ for $n \geq 1$, we again use the definition in (3.1.6) to set up the notation needed in this section.

Let $f(\zeta) = \sum_{j\geq 1} a_j\zeta^j \in \zeta\mathbf{C}[[\zeta]]$ and $u \in \underline{\mathbf{g}}$. Then set

$$\overline{u}^+(f(t^n\zeta)) = \sum_{j\geq 1} a_j\overline{u}^+_{[nj]}\otimes t^{nj}\zeta^j \tag{3.7.1}$$

(cf. (3.1.6) and (3.5.1)). We note that for $u \in \underline{\mathbf{g}}$ such that $\hat{\nu}u \neq u$,

$$\overline{u}^+(f(t^n\zeta)) = \begin{cases} \overline{u}(f(t^n\zeta)) - \overline{\hat{\nu}u}(f((-t)^n\zeta)) & \text{if } u = x_a \\ u(f(t^n\zeta)) + \hat{\nu}u(f((-t)^n\zeta)) & \text{if } u = \alpha. \end{cases} \tag{3.7.2}$$

Corollary 3.7.1 *Let $u = x_a$ and $v = x_b$ be elements of $\underline{\mathbf{g}}$ such that $\bar{a} \in \Delta^1$ and $\langle\bar{a},\bar{b}\rangle = -2$. Then for m, $n \in \mathbf{Z}$, we have*

$$\begin{aligned}&\exp\left(\overline{v}^+(t^n\zeta_2)\right)\exp\left(\overline{u}^+(t^m\zeta_1)\right) = \\ &\quad = \exp\left(\overline{u}^+(t^m\zeta_1(1+t^{m+n}\zeta_1\zeta_2)^{-1})\right)\exp\left(-([u,v])^+(\log(1+t^{m+n}\zeta_1\zeta_2))\right) \times\end{aligned}$$

$$\exp\left(\langle \overline{v}^+_{[n]}, \overline{u}^+_{[m]}\rangle n\delta_{m+n,\,0}\log(1+\zeta_1\zeta_2)c\right) \times$$
$$\exp\left(\overline{v}^+(t^n\zeta_2(1+t^{m+n}\zeta_1\zeta_2)^{-1})\right).$$

Proof: Apply (3.7.2), (3.1.2), and Proposition 3.2.1. □

Corollary 3.7.2 *Let $\bar{a} \in \Delta_{-1}$ be such that $\langle \bar{a}, \nu\bar{a}\rangle = -1$. Set $u = x_a$, $v = x_{a^{-1}}$, $w = x_{a(\not{\!\nu} a)}$ and $z = x_{(\not{\!\nu} a^{-1})a^{-1}}$. Then for m, $n \in \mathbf{Z}$ with n odd, we have*

$$\begin{aligned}
&\exp\left(\overline{z}^+(t^n\zeta_2)\right)\exp\left(\overline{u}^+(t^m\zeta_1)\right) = \\
&= \exp\left(\overline{u}^+(t^m\zeta_1(1+(-1)^m t^{2m+n}{\zeta_1}^2\zeta_2)^{-1})\right) \times \\
&\quad \exp\left(\overline{w}^+(-t^{4m+n}{\zeta_1}^4\zeta_2(1-t^{4m+2n}{\zeta_1}^4{\zeta_2}^2)^{-1})\right) \times \\
&\quad \exp\left(-([u,v])^+(\log(1+(-1)^m t^{2m+n}{\zeta_1}^2\zeta_2))\right) \times \\
&\quad \exp\left(\overline{v}^+((-1)^m t^{m+n}\zeta_1\zeta_2(1+(-1)^m t^{2m+n}{\zeta_1}^2\zeta_2)^{-1})\right) \times \\
&\quad \exp\left(\overline{z}^+(t^n\zeta_2(1-t^{4m+2n}{\zeta_1}^4{\zeta_2}^2)^{-1})\right).
\end{aligned}$$

Proof: Apply Proposition 3.2.2, (3.1.2), Proposition 3.2.1, and B-C-H to obtain the result. □

Corollary 3.7.3 *Let u, v, w, z, m and n be as in Corollary 3.7.2 with the exception that $m+n \in 2\mathbf{Z}$. Then we have*

$$\begin{aligned}
&\exp\left(\overline{v}^+(t^n\zeta_2)\right)\exp\left(\overline{u}^+(t^m\zeta_1)\right) = \\
&= \exp\left(\overline{u}^+(t^m\zeta_1(1+t^{m+n}\zeta_1\zeta_2)^{-1})\right)\exp\left(-2([u,v])^+(\log(1+t^{m+n}\zeta_1\zeta_2))\right) \times \\
&\quad \exp\left(4n\delta_{m+n,\,0}\log(1+\zeta_1\zeta_2)c\right)\exp\left(\overline{v}^+(t^n\zeta_2(1+t^{m+n}\zeta_1\zeta_2)^{-1})\right).
\end{aligned}$$

Proof: Apply Proposition 3.2.1, B-C-H, Proposition 3.2.2 and (3.1.2) to obtain the result. □

Corollary 3.7.4 *Let u, v, w, z, m, and n be as in Corollary 3.7.2 and Corollary 3.7.3 with the exception that $m+n \in 2\mathbf{Z}+1$. Then we have*

$$\exp\left(\overline{u}^+(t^n\zeta_2)\right)\exp\left(\overline{v}^+(t^m\zeta_1)\right) =$$

$$= \exp\left(\overline{u}^{+}(t^{m}\zeta_1(1 - t^{m+n}\zeta_1\zeta_2)q(t^{m+n}\zeta_1\zeta_2))\right) \times$$
$$\exp\left(\overline{w}^{+}(2(-1)^{m+1}t^{3m+n}\zeta_1{}^{3}\zeta_2 q(t^{m+n}\zeta_1\zeta_2)q((-t)^{m+n}\zeta_1\zeta_2))\right) \times$$
$$\exp\left(([u,v])\log(q(t^{m+n}\zeta_1\zeta_2))\right) \times$$
$$\exp\left(\overline{z}^{+}(2(-1)^{n+1}t^{m+3n}\zeta_1\zeta_2{}^{3}q(t^{m+n}\zeta_1\zeta_2)q((-t)^{m+n}\zeta_1\zeta_2))\right) \times$$
$$\exp\left(\overline{v}^{+}(t^{n}\zeta_2(1 - t^{m+n}\zeta_1\zeta_2)q(t^{m+n}\zeta_1\zeta_2))\right),$$

where $q(\zeta) = 1 + 2\zeta - \zeta^2$.

Proof: Apply Propositon 3.2.1, B-C-H, Proposition 3.2.2 and (3.1.2) to obtain the result. □

The other exponential identities for $A_{2n}^{(2)}$ are analogous to those in Corollaries 3.5.2 - 3.5.5.

Theorem 3.1.6 now follows for $A_{2n}^{(2)}$.

Chapter 4

Vertex algebras and integral forms of the universal enveloping algebras of the affine Lie algebras

4.1 Vertex operator algebras and vertex algebras

In the remaining sections we shift our focus to the vertex operator representations of the affine Lie algebras of the earlier sections. We will discuss and contrast two types of vertex operator descriptions of an integral form for the universal enveloping algebra $U(\underline{\ell})$.

We begin by presenting material which will be essential to the remaining sections of this paper. For more details see [F-L-M].

We will use the definition found in [F-L-M] for a vertex operator algebra:

a *vertex operator algebra* is a $\mathbf{Z}$-graded vector space

$$V = \coprod_{n \in \mathbf{Z}} V_{(n)}; \text{ for } v \in V_{(n)}, n = \text{wt } v; \tag{4.1.1}$$

such that

$$\dim V_{(n)} < \infty \text{ for } n \in \mathbf{Z} \text{ and} \tag{4.1.2}$$

$$V_{(n)} = 0 \text{ for } n \text{ sufficiently small,} \tag{4.1.3}$$

equipped with a linear map

$$\begin{aligned} V &\longrightarrow (End\ V)[[z, z^{-1}]] \\ v &\longmapsto Y(v,z) = \sum_{n \in \mathbf{Z}} v_n z^{-n-1} \text{ (where } v_n \in End\ V) \end{aligned} \tag{4.1.4}$$

and with two distinguished homogeneous vectors $\mathbf{1}, \omega \in V$, satisfying the following conditions for $u, v \in V$:

$$u_n v = 0 \text{ for } n \text{ sufficiently large;} \tag{4.1.5}$$

$$Y(\mathbf{1}, z) = 1; \tag{4.1.6}$$

$$Y(v, z)\mathbf{1} \in V[[z]] \text{ and } \lim_{z\to 0} Y(v, z)\mathbf{1} = v; \tag{4.1.7}$$

$$z_0^{-1}\delta\left(\frac{z_1 - z_2}{z_0}\right) Y(u, z_1)Y(v, z_2) - z_0^{-1}\delta\left(\frac{z_2 - z_1}{-z_0}\right) Y(v, z_2)Y(u, z_1)$$
$$= z_2^{-1}\delta\left(\tfrac{z_1 - z_0}{z_2}\right) Y(Y(u, z_0)v, z_2) \tag{4.1.8}$$

(the Jacobi identity) where $\delta(z) = \sum_{n\in\mathbf{Z}} z^n$ and where $\delta((z_1 - z_2)/z_0)$ is to be expanded as a formal power series in the second term in the numerator, z_2, and analogously for the other δ-function expressions; when each expression in (4.1.8) is applied to any element of V, the coefficient of each monomial in the formal variables is a finite sum;

$$[L(m), L(n)] = (m - n)L(m + n) + \frac{1}{12}(m^3 - m)\delta_{m+n,0}(rank\ V) \tag{4.1.9}$$

for $m, n \in \mathbf{Z}$, where

$$L(n) = \omega_{n+1} \text{ for } n \in \mathbf{Z}, \text{ i.e., } Y(\omega, z) = \sum_{n\in\mathbf{Z}} L(n)z^{-n-2} \tag{4.1.10}$$

and

$$\text{rank } V \in \mathbf{Q} \tag{4.1.11}$$

$$L(0)v = nv = (\text{wt}v)v \text{ for } n \in \mathbf{Z} \text{ and } v \in V_{(n)}; \tag{4.1.12}$$

$$\tfrac{d}{dz}Y(v, z) = Y(L(-1)v, z). \tag{4.1.13}$$

The following properties are consequences of the above definition:

$$[L(-1), Y(v, z)] = Y(L(-1)v, z); \tag{4.1.14}$$

$$[L(0), Y(v, z)] = Y(L(0)v, z) + zY(L(-1)v, z); \tag{4.1.15}$$

$$L(n)\mathbf{1} = 0 \text{ for } n \geq -1; \tag{4.1.16}$$

$$L(-2)\mathbf{1} = \omega; \tag{4.1.17}$$

$$L(0)\omega = 2\omega. \tag{4.1.18}$$

The most standard class of examples of vertex operator algebras is the class $(V_L, Y(\cdot, z))$ where V_L is a basic module for a simply-laced affine Lie algebra and $Y(v, z)$ is the general (untwisted) vertex operator associated to v in V_L. (See [F-L-M] and construction below.) A more complicated example is the Moonshine module $V^\natural$ constructed in [F-L-M]. In physics vertex operator algebras are known as chiral algebras and the associated vertex operators are examples of "quantum fields" (cf. [F-L-M]).

In this exposition we will utilize the vertex operator representations of affine Lie algebras obtained from the vertex operator algebras given by the basic modules V_L mentioned above. We will also use a slightly weaker structure than a vertex operator algebra, called a *vertex algebra*, to examine the standard representations of the affine Lie algebras of level $k \geq 1$. A *vertex algebra* V is a $\mathbf{Z}$-graded vector space satisfying (4.1.1) equipped with a linear map as in (4.1.4) such that conditions (4.1.5) through (4.1.13) hold for $u, v \in V$ (and hence through (4.1.18)). Vertex algebras were first studied by Borcherds in his announcement [B]. Unlike Borcherds' definition in [B], our concept of a vertex algebra includes the Jacobi identity (condition (4.1.8)) and properties concerning the Virasoro algebra. We also note that Borcherds uses components as opposed to the generating functions utilized in [F-L-M].

The vertex algebras considered here are constructed, following [B] and [F-L-M], in the same manner as the basic modules V_L. We use even nondegenerate lattices L to construct the vertex algebras V_L. If the lattice L is an even positive definite lattice, we see that the vertex algebra V_L satisfies conditions (4.1.2) and (4.1.3), and so is a vertex operator algebra.

We now turn to a brief discussion which describes the construction of the vertex algebra V_L associated with an even nondegenerate lattice L. For more details see Chapters 5, 6, 7, and 8 of [F-L-M].

Let L be an even nondegenerate lattice of rank l ($l \geq 0$), with a symmetric bilinear form $\langle \cdot, \cdot \rangle : L \times L \longrightarrow \mathbf{Z}$. Note that we do not require L to be positive definite (cf.

§2.1). Suppose further that L is spanned (over $\mathbf{Z}$) by the set

$$L_2 = \{\alpha \in L | \langle \alpha, \alpha \rangle = 2\}, \tag{4.1.19}$$

such that the set $\Pi = \{\alpha_1, \alpha_2, \ldots, \alpha_l\}$ ($\subset L_2$) is a ($\mathbf{Z}$-)basis of L (cf. (2.1.1)). As in §2.1 form the Lie algebras $\underline{h} = L \otimes_{\mathbf{Z}} \mathbf{C}$ and $\tilde{\underline{h}}$, and extend $\langle \cdot, \cdot \rangle$ to each of these algebras in the natural way. Recall that the form $\langle \cdot, \cdot \rangle$ is symmetric, bilinear, and invariant on each of the respective algebras.

Let $\hat{\underline{h}}_{\mathbf{Z}}$ denote the commutator subalgebra of $\tilde{\underline{h}}$, i.e., $\hat{\underline{h}}_{\mathbf{Z}} = \tilde{\underline{h}}'$. Using (2.1.11) and (2.1.12) we see that $\hat{\underline{h}}_{\mathbf{Z}} = \coprod_{n \in \pm \mathbf{Z}_+} \underline{h} \otimes t^n \oplus \mathbf{C}c$. Next set

$$\hat{\underline{h}}_{\mathbf{Z}}^{-} = \coprod_{n<0} \underline{h} \otimes t^n, \; \hat{\underline{h}}_{\mathbf{Z}}^{+} = \coprod_{n>0} \underline{h} \otimes t^n \text{ and } \hat{\underline{b}}_{\mathbf{Z}} = \hat{\underline{h}}_{\mathbf{Z}}^{+} \oplus \mathbf{C}c.$$

Let $h(n)$ denote the element $h \otimes t^n$ for $h \in \underline{h}$, $n \in \mathbf{Z}$. Then form the irreducible induced $\hat{\underline{h}}_{\mathbf{Z}}$-module given by

$$M(1) = \mathrm{Ind}_{\mathrm{U}(\hat{\underline{b}}_{\mathbf{Z}})}^{\mathrm{U}(\hat{\underline{h}}_{\mathbf{Z}})} \mathbf{C}_1 = \mathrm{U}(\hat{\underline{h}}_{\mathbf{Z}}) \otimes_{\mathrm{U}(\hat{\underline{b}}_{\mathbf{Z}})} \mathbf{C}_1 \cong \mathrm{U}(\hat{\underline{h}}_{\mathbf{Z}}^{-}) \otimes \mathbf{C}_1$$

where $\hat{\underline{h}}_{\mathbf{Z}}^{+}$ acts trivially on the 1-dimensional module $\mathbf{C}_1$ and the central element c acts as 1 on $\mathbf{C}_1$. Note that since $\hat{\underline{h}}_{\mathbf{Z}}^{-}$ is abelian, $M(1)$ is isomorphic to $\mathbf{S}(\hat{\underline{h}}_{\mathbf{Z}}^{-}) \otimes \mathbf{C}_1$. $M(1)$ is also an $\tilde{\underline{h}}$-module by defining $h(0)$ ($h \in \underline{h}$) to act trivially on $\mathbf{C}_1$ and d to act as $d \otimes 1$. The module $M(1)$ is irreducible as both an $\hat{\underline{h}}_{\mathbf{Z}}^{+}$-module and as an $\tilde{\underline{h}}$-module (cf. [F-L-M]).

Let $(\hat{L}, -)$ and c_0 be as in (2.1.2) and (2.1.3). Again as in §2.1, assume that $c_0(\bar{a}, \bar{b}) = \langle \bar{a}, \bar{b} \rangle \bmod 2\mathbf{Z}$. Next form the induced $\hat{L}$-module

$$\mathbf{C}\{L\} = \mathrm{Ind}_{\mathbf{C}[\langle\kappa\rangle]}^{\mathbf{C}[\hat{L}]} \mathbf{C}_\chi \cong \mathbf{C}[\hat{L}] \otimes_{\mathbf{C}[<\kappa>]} \mathbf{C}_\chi \cong \mathbf{C}[\hat{L}]/(1+\kappa)\mathbf{C}[\hat{L}] \overset{\text{(linearly)}}{\cong} \mathbf{C}[\mathrm{L}], \tag{4.1.20}$$

where $\chi : \langle \kappa \rangle \longrightarrow \mathbf{C}^\times$ is a faithful character with $\chi(\kappa) = -1$. Let $\iota(a)$ denote the element $a \otimes 1 = \iota(a)$ in $\mathbf{C}\{L\}$ for $a \in \hat{L}$. It is easily seen that the elements $\iota(a)$ span $\mathbf{C}\{L\}$, and using (4.1.20), we see that the action of $\hat{L}$ on $\mathbf{C}\{L\}$ is given by

$$\begin{aligned} a \cdot \iota(b) &= \iota(ab) \\ \kappa \cdot \iota(b) &= \iota(\kappa b) = -\iota(b) \end{aligned} \tag{4.1.21}$$

where a, b, $\kappa \in \hat{L}$.

Define $deg\ \iota(a) = -\frac{1}{2}\langle \overline{a}, \overline{a}\rangle + \frac{1}{24} dim\ \underline{\mathrm{h}}$ for $a \in \hat{L}$. Then define an action of $\underline{\mathrm{h}}$ on $\mathbf{C}\{L\}$ by:

$$h(0) \cdot \iota(a) = \langle h, \overline{a}\rangle \iota(a) \tag{4.1.22}$$

for $a \in \hat{L}$ and $h \in \underline{\mathrm{h}}$. $\mathbf{C}\{L\}$ also may be considered an $\tilde{\underline{\mathrm{h}}}$-module by allowing $\underline{\mathrm{h}}$ to act trivially, $\underline{\mathrm{h}}(0)$ to act as in (4.1.22) and the action of d to be given by

$$d \cdot \iota(a) = (\deg\ \iota(a))\iota(a). \tag{4.1.23}$$

Set $V_L = \mathbf{S}(\hat{\underline{\mathrm{h}}}_{\mathbf{Z}}^-) \otimes \mathbf{C}\{L\}$. Since both $\mathbf{S}(\hat{\underline{\mathrm{h}}}_{\mathbf{Z}}^-)$ and $\mathbf{C}\{L\}$ are $\tilde{\underline{\mathrm{h}}}$-modules, we have V_L is an $\tilde{\underline{\mathrm{h}}}$-module via the tensor product action. We may also view $\mathbf{S}(\hat{\underline{\mathrm{h}}}_{\mathbf{Z}}^-)$ as a trivial $\hat{L}$-module. Thus V_L is also an $\hat{L}$-module via the tensor product action.

We next describe another operator on the space V_L. Let z, z_1, z_2, etc. be independent commutating formal variables. Define an action of $z^{\alpha(0)} = z^{\alpha}$ on $\mathbf{C}\{L\}[[z, z^{-1}]]$ by

$$z^{\alpha(0)} \cdot \iota(a) = z^{\alpha} \cdot \iota(a) = z^{\langle \overline{a}, \alpha\rangle}\iota(a) \tag{4.1.24}$$

for $a \in \hat{L}, \alpha \in L$. Allow z^{α} to act trivially on $\mathbf{S}(\hat{\underline{\mathrm{h}}}_{\mathbf{Z}}^-)$. Thus there is an action of z^{α} on V_L given by $z^{\alpha} \mapsto 1 \otimes z^{\alpha}$.

These elements can be used to create formal Laurent power series with coefficients in $(End\ V_L)$ which act on $V_L[[z, z^{-1}]]$. For $\alpha \in L$ define

$$E^{\pm}(\alpha, z) = \exp\left(\textstyle\sum_{n>0} \frac{\alpha(\pm n)}{\pm n} z^{\mp n}\right) \tag{4.1.25}$$

$$\alpha(z) = \textstyle\sum_{n \in \mathbf{Z}} \alpha(n) z^{-n-1} \tag{4.1.26}$$

$$\alpha(z)^- = \textstyle\sum_{n<0} \alpha(n) z^{-n-1} \tag{4.1.27}$$

$$\alpha(z)^+ = \textstyle\sum_{n\geq 0} \alpha(n) z^{-n-1} \tag{4.1.28}$$

in $(End\ V_L)[[z, z^{-1}]]$. Using (4.1.21), (4.1.24) and (4.1.25) define the *(untwisted) vertex operator* associated to $a \in \hat{L}$ [F-L-M] by

$$Y(a, z) = Y(\iota(a), z) = E^-(-\overline{a}, z)E^+(-\overline{a}, z)az^{\overline{a}} = \sum_{n \in \mathbf{Z}} \iota(a)_n z^{-n-1}. \tag{4.1.29}$$

To fulfill (4.1.4) we need to extend the definition of a vertex operator to each element $v \in V_L$. In order to accomplish this, we will have to multiply some of the above formal Laurent series together. However arbitrary products of this nature are not necessarily well-defined: the coefficient of each power of z may not be summable, i.e., a finite expression in $End\ V_L$. This problem is alleviated by introducing the notion of a *normal ordered product.* Roughly speaking, *normal ordering* is the placement of annihilation operators to the right of creation operators. More precisely, we define the following normal ordered products in $End\ V_L$:

$$\begin{aligned}
{}^{\circ}_{\circ}\alpha(z_1)\beta(z_2){}^{\circ}_{\circ} &= {}^{\circ}_{\circ}\beta(z_2)\alpha(z_1){}^{\circ}_{\circ} \\
&= \alpha(z_1)^-\beta(z_2)^- + \alpha(z_1)^-\beta(z_2)^+ \\
&\quad +\beta(z_2)^-\alpha(z_1)^+ + \alpha(z_1)^+\beta(z_2)^+ \\
{}^{\circ}_{\circ}\alpha(0)a{}^{\circ}_{\circ} &= {}^{\circ}_{\circ}a\alpha(0){}^{\circ}_{\circ} = a\alpha(0) \qquad (4.1.30) \\
{}^{\circ}_{\circ}\alpha(0)z^{\overline{a}}{}^{\circ}_{\circ} &= {}^{\circ}_{\circ}z^{\overline{a}}\alpha(0){}^{\circ}_{\circ} = \alpha(0)z^{\overline{a}} \\
{}^{\circ}_{\circ}\alpha(z_1)Y(a,z_2){}^{\circ}_{\circ} &= {}^{\circ}_{\circ}Y(a,z_2)\alpha(z_1){}^{\circ}_{\circ} \\
&= \alpha(z_1)^-Y(a,z_2) + Y(a,z_2)\alpha(z_1)^+
\end{aligned}$$

where $\alpha \in L$ and $a \in \hat{L}$. Recall $Y(\cdot, z)$ is to be a linear map. So to extend the definition of a vertex operator to all of V_L, it is enough to define $Y(\cdot, z)$ for $v \in V_L$ of the form

$$v = \alpha_1(-n_1)\alpha_2(-n_2)\cdots\alpha_k(-n_k) \otimes \iota(a) \qquad (4.1.31)$$

where $\alpha_i \in L, n_i \in \mathbf{Z}_+$, and $a \in \hat{L}$. Define the *general (untwisted) vertex operator*, $Y(v, z)$, associated to $v \in V_L$ ([F-L-M], [B]) by

$$\begin{aligned}
Y(v,z) &= {}^{\circ}_{\circ}\left[\frac{1}{(n_1-1)!}\left(\frac{d}{dz}\right)^{n_1-1}\alpha_1(z)\right]\cdots\left[\frac{1}{(n_k-1)!}\left(\frac{d}{dz}\right)^{n_k-1}\alpha_k(z)\right]Y(a,z){}^{\circ}_{\circ} \\
&= \sum_{n\in\mathbf{Z}} v_n z^{-n-1} \qquad (4.1.32)
\end{aligned}$$

We list some immediate consequences of (4.1.32):

1. $Y(1,z) = Y(\iota(1),z) = 1_{|V_L} = \iota(1)_{-1}$.

2. $Y(\alpha(-1)\otimes\iota(1),z) = {}^{\circ}_{\circ}\alpha(z)Y(1,z){}^{\circ}_{\circ} = {}^{\circ}_{\circ}\alpha(z){}^{\circ}_{\circ} = \alpha(z)$ for $\alpha \in L$.

3. Definitions (4.1.29) and (4.1.32) coincide for $a \in \hat{L}$.

It is also seen (although not immediately!) that the Jacobi identity holds for $u,\ v \in V_L$ (cf. §8.8 of [F-L-M]).

Theorem 4.1.1 ([B], [F-L-M]) *For $u,\ v \in V_L$, we have*

$$z_0^{-1}\delta\left(\frac{z_1 - z_2}{z_0}\right)Y(u,z_1)Y(v,z_2) - z_0^{-1}\delta\left(\frac{z_2 - z_1}{-z_0}\right)Y(v,z_2)Y(u,z_1)$$
$$= z_2^{-1}\delta\left(\tfrac{z_1-z_0}{z_2}\right)Y(Y(u,z_0)v,z_2)$$

We next state two useful corollaries to the above theorem.

Corollary 4.1.2 ([B]) *For $u, v \in V_L$ the general (untwisted) vertex operators associated to u and v, respectively, have the following commutator formula:*

$$[Y(u,z_1),Y(v,z_2)] = \mathrm{Res}_{z_0} z_2^{-1}Y(Y(u,z_0)v,z_2)e^{-z_0\,\partial/\partial z_1}\delta(z_1/z_2). \qquad (4.1.33)$$

Proof: See [F-L-M]; Borcherds formulates this result in component form. □

Corollary 4.1.3 ([B]) *Let $u, v \in V_L$ and $n, m \in \mathbf{Z}$. Then*

$$[u_m, v_n] = \sum_{i\in\mathbf{N}} \binom{m}{i} (u_i \cdot v)_{m+n-i}, \qquad (4.1.34)$$

where $u_i \cdot v$ denotes the image of v under $u_i \in End\, V_L$, i.e., the coefficient of z^{-i-1} in $Y(u,z)\cdot v$.

Let $\{h_1, h_2, \ldots, h_l\}$ be an orthonormal basis of $\underline{h}$. Define the distinguished element ω of V_L (cf. [F-L-M]) by

$$\omega = \sum_{i=1}^{l}(h_i(-1))^2 \otimes \iota(1). \qquad (4.1.35)$$

Then using (4.1.32) $Y(\omega,z)$ can be written as

$$Y(\omega,z) = \sum_{n\in\mathbf{Z}} \omega_n z^{-n-1} = \sum_{n\in\mathbf{Z}} L(n) z^{-n-2},$$

where the component operators $L(n)$ are given by

$$L(n) = \frac{1}{2}\sum_{i=1}^{l}\left(\sum_{k\in\mathbf{Z}} {}^{\circ}_{\circ} h_i(n-k)h_i(k) {}^{\circ}_{\circ}\right). \tag{4.1.36}$$

Using Corollary 4.1.3 and (4.1.36) we have

$$[L(m), L(n)] = (m-n)L(m+n) + \frac{1}{12}(m^3 - m)\delta_{m+n,\,0}(dim\, \underline{\mathrm{h}}).$$

Thus the component operators $L(n)$ along with $(rank\ V_L)\iota(1)_{-1}$ give a representation of a Virasoro algebra (cf. [F-L-M]).

Let $v \in V_L$ be a homogeneous vector given by (4.1.31) such that $deg\ v = -n + \frac{1}{24}dim\, \underline{\mathrm{h}}$. Then we see using (4.1.36), that

$$L(0)\cdot v = -n_1 - n_2 - \cdots - n_k - \frac{1}{2}\langle\overline{a}, \overline{a}\rangle + \frac{1}{24}dim\, \underline{\mathrm{h}} = nv + \frac{1}{24}dim\, \underline{\mathrm{h}}.$$

Define wt v to be the $L(0)$-eigenvalue for the vector v, i.e., wt $v = n$. Note that $d = -L(0) + \frac{1}{24}dim\, \underline{\mathrm{h}}$.

Theorem 4.1.4 ([B], [F-L-M]) *For a positive definite even lattice L, the space V_L based on the central extension of L by $\langle\pm 1\rangle$ with commutator map given by $c_0(\alpha,\beta) = \langle\alpha,\beta\rangle \bmod 2\mathbf{Z}$ $(\alpha,\beta \in L)$ has the structure of a vertex operator algebra with rank V_L = rank L.*

Proof: See Chapter 8 in [F-L-M]. □

Theorem 4.1.5 ([B], [F-L-M]) *Let L be a nondegenerate even lattice. Then the space V_L based on the central extension of L by $\langle\pm 1\rangle$ with commutator map given by $c_0(\alpha,\beta) = \langle\alpha,\beta\rangle \bmod 2\mathbf{Z}$ $(\alpha,\beta \in L)$ is a vertex algebra.*

4.2 Vertex operator representations of $A_l^{(1)}$ $(l \geq 1)$, $D_l^{(1)}$ $(l \geq 1)$, $E_l^{(1)}$ $(l = 6,\ 7,\ 8)$

Let L be the root lattice associated to Δ with Δ as in §2.1. Construct the vertex operator algebra V_L (cf. §4.1). The aim of this section is to realize the affines $\tilde{\mathfrak{g}}$, $\tilde{\mathfrak{g}}_{[0]}$ and $\tilde{\mathfrak{g}}^{(\tau)}$ in terms of the component opperators of certain elements in V_L. We start by recalling two properties concerning the component operators of the vertex operators $Y(u,z)$ from [F-L-M].

Proposition 4.2.1 *Let u and v be two homogeneous elements of V_L. Then the element $u_n \cdot v$ $(n \in \mathbf{Z})$ has*

$$\mathrm{wt}\ (u_n \cdot v) = \mathrm{wt}\ u + \mathrm{wt}\ v - n - 1. \tag{4.2.1}$$

In particular, if $\mathrm{wt}\ u = \mathrm{wt}\ v = 1$, *then* $\mathrm{wt}\ (u_0 \cdot v) = 1$ *and* $\mathrm{wt}\ (u_1 \cdot v) = 0$.

Proof: To determine the weight of the element $u_n \cdot v$ we compute $L(0) \cdot (u_n \cdot v)$. Using (4.1.12) - (4.1.15) we see that

$$L(0) \cdot Y(u,z)v = (\mathrm{wt}\ v)Y(u,z)v + (\mathrm{wt}\ u)Y(u,z)v + z\frac{d}{dz}Y(u,z)v. \tag{4.2.2}$$

The result is then read off as the coefficient of z^{-n-1} in (4.2.2). □

The next proposition gives us a commutator formula for component operators associated to elements of of V_L of weight one.

Proposition 4.2.2 *Let u and v be elements in V_L such that* $\mathrm{wt}\ u = \mathrm{wt}\ v = 1$. *Then*

$$[u_m, v_n] = (u_0 \cdot v)_{m+n} + m(u_1 \cdot v)\delta_{m+n,\,0}.$$

Proof: By Corollary 4.1.3, $[u_m, v_n] = \sum_{i \in \mathbf{N}} \binom{m}{i} (u_i \cdot v)_{m+n-i}$. Since L is a positive definite lattice, we have $(V_L)_{(n)} = 0$ for $n < 0$. By Proposition 4.2.1 $\mathrm{wt}(u_i \cdot v) = 1 - i$. Thus the sum has at most two terms, and the element $(u_1 \cdot v)$ is of weight zero which implies $(u_1 \cdot v)$ is a multiple of $\iota(1)$, since $\iota(1)_j = \delta_{-1,j}$, (cf. (4.1.4), (4.1.6)). □

The preceding proposition shows that the space spanned by the component operators of the elements of weight one along with $\iota(1)_{-1}$ is closed under the bracket operation. The next step in realizing the affines in terms of vertex operators is to see how the different component operators bracket. We consider the elements of weight one in V_L. These vectors are linear combinations of elements of the form

$$\{\iota(a) \mid \overline{a} \in \Delta\} \cup \{\alpha(-1) \otimes \iota(1) \mid \alpha \in L\} \tag{4.2.3}$$

(cf. [F-L-M]). Since $Y(\cdot, z)$ is linear, it is sufficient to examine the bracket relations of the operators associated to u and v in $(V_L)_{(1)}$ with u and v given by $\iota(a)$ or $\alpha(-1) \otimes \iota(1)$ in (4.2.3).

Proposition 4.2.3 *(1) Let* $u = \alpha(-1)\otimes\iota(1)$, $v = \beta(-1)\otimes\iota(1)$, *and* $m, n \in \mathbf{Z}$, *where* $\alpha, \beta \in L$. *Then*

$$[u_m, v_n] = m\langle\alpha, \beta\rangle\delta_{m+n,\,0}. \tag{4.2.4}$$

(2) Let $u = \alpha(-1) \otimes \iota(1)$, $v = \iota(a)$ *and* $m, n \in \mathbf{Z}$ *with* $\alpha \in L$ *and* $b \in \hat{\Delta}$. *Then*

$$[u_m, v_n] = \langle\alpha, \overline{b}\rangle\iota(b)_{m+n}. \tag{4.2.5}$$

(3) Let $u = \iota(a)$, $v = \iota(b)$ *and* $m, n \in \mathbf{Z}$ *with* $a, b \in \hat{\Delta}$. *Then*

$$[u_m, v_n] = \begin{cases} \overline{a}(-1)_{m+n} + m\iota(1)_{-1}\delta_{m+n,\,0} & \textit{if} \quad b = a^{-1} \\ \iota(ab)_{m+n} & \textit{if} \quad \langle\overline{a}, \overline{b}\rangle = -1 \\ 0 & \textit{if} \quad \langle\overline{a}, \overline{b}\rangle \geq 0. \end{cases} \tag{4.2.6}$$

(4) Let $u \in (V_L)_{(1)}$ *and* $n \in \mathbf{Z}$. *Then*

$$[d, u_n] = [-L(0) + \frac{1}{24} dim\ \underline{h}\ \iota(1)_{-1}, u_n] = nu_n. \tag{4.2.7}$$

Proof: Use Corollary 4.1.3 and (4.1.36) (cf. [F-L-M]). □

The preceding propositions show that the component operators of the homogeneous elements of weight one, along with $-L(0) + \frac{1}{24} dim\ \underline{h}\ \iota(1)_{-1}$, generate a Lie algebra $\underline{\ell}$. In fact we have:

Theorem 4.2.4 ([F-K], [Seg]) *The map*

$$\varphi : \underline{\tilde{g}} \longrightarrow (End\, V_L)[[z, z^{-1}]]$$

given by

$$\begin{aligned} x_a \otimes t^n &\longmapsto \iota(a)_n \\ \alpha \otimes t^n &\longmapsto (\alpha(-1) \otimes \iota(1))_n = \alpha(-1)_n \\ c &\longmapsto \iota(1)_{-1} \\ d &\longmapsto -L(0) + \frac{1}{24} dim\, \underline{h}\, \iota(1)_{-1} \end{aligned}$$

is a representation of the Lie algebra $\underline{\tilde{g}}$. Furthermore, φ is an isomorphism between $\underline{\tilde{g}}$ and the Lie algebra $\underline{\ell}$ generated by the component operators of the elements of weight one and the degree operator $-L(0) + \frac{1}{24} dim\, \underline{h}\, \iota(1)_{-1}$.

Recall that the affine Lie algebras in §2.1 were equipped with a symmetric bilinear $\underline{\tilde{g}}$-invariant form $\langle \cdot, \cdot \rangle$. To build such a form on $\underline{\ell}$ we start with the form $\langle \cdot, \cdot \rangle$ on $\underline{h} \times \underline{h}$. Extend $\langle \cdot, \cdot \rangle$ to $\mathbf{C}\{L\}_{(1)} \times \mathbf{C}\{L\}_{(1)}$ by defining

$$\begin{aligned} \langle \iota(a), \iota(b) \rangle &= \begin{cases} 1 & if \quad ab = 1 \\ 0 & if \quad ab \notin \{1,\ \kappa\} \end{cases} \\ \langle \alpha(-1) \otimes \iota(1), \beta(-1) \otimes \iota(1) \rangle &= \langle \alpha, \beta \rangle \\ \langle \alpha(-1) \otimes \iota(1), \iota(b) \rangle &= \langle \iota(b), \alpha(-1) \otimes \iota(1) \rangle = 0 \end{aligned} \tag{4.2.8}$$

Next extend $\langle \cdot, \cdot \rangle$ to $\underline{\ell} \times \underline{\ell}$ by

$$\begin{aligned} \langle u_m, v_n \rangle &= \langle u, v \rangle \delta_{m+n,\,0} \\ \langle \iota(1)_{-1}, v_n \rangle &= \langle v_n, \iota(1)_{-1} \rangle = 0 \\ \langle L(0), v_n \rangle &= \langle v_n, L(0) \rangle = 0 \\ \langle \iota(1)_{-1}, L(0) \rangle &= \langle L(0), \iota(1)_{-1} \rangle = -1 \\ \langle \iota(1)_{-1}, \iota(1)_{-1} \rangle &= \langle L(0), L(0) \rangle = 0 \end{aligned} \tag{4.2.9}$$

where $u,\ v \in (V_L)_{(1)}$ and $m,\ n \in \mathbf{Z}$. This gives a $\underline{\ell}$-invariant form on $\underline{\ell} \times \underline{\ell}$.

Using φ and the $\underline{\ell}$-invariant form $\langle\cdot,\cdot\rangle$ we may rewrite §2.1 in terms of the vertex operator representation. So the Cartan subalgebra $\varphi(\underline{\mathrm{h}}^e)$ becomes

$$(\underline{\mathrm{h}}(-1))_0 \oplus \mathbf{C}\iota(1)_{-1} \oplus \mathbf{C}(-L(0) + \frac{1}{24} dim\,\underline{\mathrm{h}}\,\iota(1)_{-1}), \tag{4.2.10}$$

and the root system $\varphi(\Delta(\tilde{\underline{\mathrm{g}}}))$ is given by

$$\{\alpha(-1)_0 + m\iota(1)_{-1} \mid \alpha \in \Delta,\ m \in \mathbf{Z}\} \cup \{n\iota(1)_{-1} \mid n \in \mathbf{Z} \setminus \{0\}\}. \tag{4.2.11}$$

We are motivated by [G] and §2.1 to give the following definition of a Chevalley basis for $\underline{\ell}$. Choose a section $e : L \to \hat{L}$. We define a *Chevalley basis* of $\underline{\ell}$ to be a basis for $\underline{\ell}$ of the form

$$\{\iota(e_\alpha)_n \mid n \in \mathbf{Z},\ \alpha \in \Delta\} \cup \{\alpha^\vee(-1)_n \mid \alpha \in \Pi,\ n \in \mathbf{Z}\}$$
$$\cup \{(\alpha_0(-1)_0 + \iota(1)_{-1})^\vee\} \cup \{-L(0) + \tfrac{1}{24} dim\,\underline{\mathrm{h}}\,\iota(1)_{-1}\}.$$

As a consequence of Proposition 4.2.3 and the fact that $\alpha^\vee = \alpha$ for $\alpha \in \Delta$, we have

$$[\iota(e_\alpha)_n, \iota(e_{-\alpha})_{-n}] = \epsilon(\alpha, -\alpha)\,(\alpha(-1)_0 + n\iota(1)_{-1})^\vee, \tag{4.2.12}$$

and the linear map $\theta : \underline{\ell} \to \underline{\ell}$ given by

$$\begin{aligned} \iota(e_\alpha)_n &\longmapsto -\epsilon(\alpha, -\alpha)\iota(e_{-\alpha})_{-n} \\ \alpha(-1)_n &\longmapsto -\alpha(-1)_{-n} \\ \iota(1)_{-1} &\longmapsto -\iota(1)_{-1} \\ L(0) &\longmapsto -L(0) \end{aligned} \tag{4.2.13}$$

where $\alpha \in \Delta$ and $n \in \mathbf{Z}$, is a Lie algebra automorphism of $\underline{\ell}$.

Remark 4.2.5 We note that within the realm of vertex operator theory, (4.2.12) and (4.2.13) are automatic. Hence our definition of a Chevalley basis always satisfies Garland's definition.

We are able to recover the vertex operator version of Theorem 2.1.2.

Theorem 4.2.6 *Choose a section $e : L \to \hat{L}$ as in Theorem 2.1.2. Then the set*

$$\begin{aligned} \mathbf{S} &= \{\iota(e_\alpha)_n \mid \alpha \in \Delta,\ n \in \mathbf{Z}\} \cup \{\alpha(-1)_n \mid \alpha \in \Pi,\ n \in \mathbf{Z}\} \\ &\cup \{\alpha_0(-1)_0 + \iota(1)_{-1}\} \cup \{-L(0) + \frac{1}{24} dim\, \underline{h}\, \iota(1)_{-1}\} \end{aligned}$$

is a Chevalley basis of $\underline{\ell}$.

We can also construct the affines $\underline{\ell}_{[0]}$ and $\underline{\ell}^{(\tau)}$ using Sections 2.2 - 2.6, φ, $\underline{\ell}$ of Theorem 4.2.4, and the $\underline{\ell}$-invariant form $\langle \cdot, \cdot \rangle$ of (4.2.9). Thus we have

Theorem 4.2.7 *Fix a linearly ordered Chevalley basis $\mathbf{S}$ of $\underline{\ell}$. Let $U_{\mathbf{Z}}(\underline{\ell})$ be the $\mathbf{Z}$-subalgebra of $U(\underline{\ell})$ generated by the elements*

$$\Lambda_s((\iota(a)^+_{[Nm],Nm})_N) \text{ and } \Lambda_s((-(-L(0) + \tfrac{1}{24} dim\, \underline{h}\, \iota(1)_{-1}))),$$

(cf. (3.1.12), (3.1.14)), where $\overline{a}_{[0]}(-1)_0 + m\iota(1)_{-1} \in \Delta$ and $s \in \mathbf{N}$. Then the set of (ordered) monomials forms an integral basis of $U_{\mathbf{Z}}$, and $U_{\mathbf{Z}}$ is an integral form of $U(\underline{\ell})$, i.e., $U_{\mathbf{Z}} \otimes_{\mathbf{Z}} \mathbf{C} = U(\underline{\ell})$.

4.3 Schur polynomials and $S(\hat{\underline{h}}^-_{\mathbf{Z}})$

Schur polynomials were used in [L-P] to prove independence of a basis for the vacuum space Ω_L for a standard $\tilde{\underline{sl}}(2)$-module L. The observations made in [L-P] gave a relationship between the generating functions $E^-(-\alpha, z)$ (cf. (4.1.25)) and the Schur polynomials in the variables $\alpha(-n)$. Using this relationship for $\tilde{\underline{sl}}(2)$, we are able to describe a basis for the symmetric algebra $S(\hat{\underline{h}}^-_{\mathbf{Z}})$ associated with even nondegenerate lattices other than the root lattice for $\underline{sl}(2)$. The notations and definitions below may be found in [Mac] or in §7 of [L-P].

Define a *partition* λ, $\lambda = (\lambda_1, \lambda_2, \ldots, \lambda_n, \ldots)$, to be a (finite or infinite) decreasing sequence of nonnegative integers

$$\lambda_1 \geq \lambda_2 \geq \cdots$$

such that only finitely many of the $\lambda_i \neq 0$. The λ_i such that $\lambda_i \neq 0$ are called the *parts* of λ. The *length* $l(\lambda)$ of a partition is the number of parts of λ. The *weight* of λ, $|\lambda|$, is given by

$$|\lambda| = \lambda_1 + \lambda_2 + \cdots + \lambda_n + \cdots.$$

For $i \in \mathbf{Z}_+$ set $m_i(\lambda) = card\{j \mid \lambda_j = i\}$; $m_i(\lambda)$ is called the *multiplicity of i in λ*.

First we examine $S(\hat{\underline{h}}_{\mathbf{Z}}^-)$ where $\underline{h} = L \otimes_{\mathbf{Z}} \mathbf{C}$ with $L = \mathbf{Z}\alpha$ and $\langle \alpha, \alpha \rangle = 2$ (cf. [L-P]). Recall that $S(\hat{\underline{h}}_{\mathbf{Z}}^-)$ is abelian, so we may identify $S(\hat{\underline{h}}_{\mathbf{Z}}^-)$ with the polynomial ring $\mathbf{C}[\alpha(-1), \alpha(-2), \ldots, \alpha(-n), \ldots]$. For $\lambda = (\lambda_1, \lambda_2, \ldots)$ a partition, let $p_{-\lambda} \in S(\hat{\underline{h}}_{\mathbf{Z}}^-)$ be given by

$$p_{-\lambda} = \begin{cases} \alpha(-\lambda_1)\alpha(-\lambda_2)\cdots & if \quad \lambda_1 \neq 0 \\ 1 & if \quad \lambda_1 = 0, \end{cases} \tag{4.3.1}$$

where each λ_i is a part of λ. We see that the family $\{p_{-\lambda} \mid \lambda \text{ is a partition}\}$ forms a basis of $S(\hat{\underline{h}}_{\mathbf{Z}}^-)$.

We may also express $p_{-\lambda}$ as the product

$$p_{-\lambda} = \prod_{i \geq 1} \alpha(-i)^{m_i(\lambda)}. \tag{4.3.2}$$

Next set

$$z_\lambda = \prod_{i \geq 1} i^{m_i(\lambda)} m_i(\lambda)!, \tag{4.3.3}$$

and define

$$h_{-n} = \begin{cases} \sum_{|\lambda|=n} z_\lambda^{-1} p_{-\lambda} & if \quad n > 0 \\ 1 & if \quad n = 0 \\ 0 & if \quad n < 0. \end{cases} \tag{4.3.4}$$

For a sequence $\lambda = (\lambda_1, \lambda_2, \ldots)$ where the $\lambda_i \geq 0$ and finitely many λ_i are nonzero, define

$$h_{-\lambda} = h_{-\lambda_1} h_{-\lambda_2} \cdots. \tag{4.3.5}$$

Proposition 4.3.1 *The set*

$$\{h_{-\lambda} \mid \lambda \text{ is a partition}\} \tag{4.3.6}$$

is a basis of $S(\hat{\underline{\mathfrak{h}}}_{\mathbf{Z}}^{-})$*. Moreover, the* **Z***-subalgebra generated by this set is a* **Z***-form of* $S(\hat{\underline{\mathfrak{h}}}_{\mathbf{Z}}^{-})$.

Proof: The fact that $\{h_{-\lambda} \mid \lambda \text{ is a partition}\}$ is a basis of $S(\hat{\underline{\mathfrak{h}}}_{\mathbf{Z}}^{-})$ follows from Proposition 7.1 of [L-P].

Since $S(\hat{\underline{\mathfrak{h}}}_{\mathbf{Z}}^{-})$ is abelian, we observe that $h_{-n}h_{-m} = h_{-m}h_{-n}$. Thus

$$h_{-\lambda}h_{-\mu} = h_{-\lambda_1}h_{-\lambda_2}\cdots h_{-\lambda_n}h_{-\mu_1}\cdots h_{-\mu_m} = h_{(\tau_1,\tau_2,\ldots,\tau_{m+n},0,\ldots)}$$

with τ a partition such that the parts of τ are made up of the parts of λ and μ, and $l(\tau) = l(\lambda) + l(\mu)$. Hence the **Z**-subalgebra generated by (4.3.6) is an integral form of $S(\hat{\underline{\mathfrak{h}}}_{\mathbf{Z}}^{-})$. □

For $n \geq 0$, let $\delta = \delta_n = (n-1, n-2, \ldots, 1, 0)$ in $\mathbf{Z}^n$. Note that S_n, the symmetric group acting on n elements, acts on $\mathbf{Z}^n$ by permutation of the coordinates. Again since $S(\hat{\underline{\mathfrak{h}}}_{\mathbf{Z}}^{-})$ is abelian, we have

$$h_{-w\lambda} = h_{-\lambda} \tag{4.3.7}$$

where $w \in S_n$ and $\lambda \in \mathbf{Z}^n$. Let $\lambda \in \mathbf{Z}^n$, and define the *Schur function (or Schur polynomial)* $s_{-\lambda}$ to be

$$s_{-\lambda} = \sum_{w \in S_n} \varepsilon(w) h_{-(\lambda+\delta-w\delta)}, \tag{4.3.8}$$

where $\varepsilon(w)$ is the sign of the permutation w.

We next record a few facts from [L-P] which will be needed later.

Lemma 4.3.2 *Let* λ *and* μ *be in* $\mathbf{Z}^n$. *Suppose there exists* $w' \in S_n$ *such that* $w'(\lambda + \delta) = \mu + \delta$. *Then* $s_{-\lambda} = \varepsilon(w')s_{-\mu}$.

Proof: This follows from the definition of s_λ and (4.3.7). □

Lemma 4.3.3 *Let* $\lambda = (\lambda_1, \lambda_2, \ldots, \lambda_n)$ *be a partition. Then*

$$s_{-(\lambda_1,\lambda_2,\ldots,\lambda_n)} = s_{-(\lambda_1,\lambda_2,\ldots,\lambda_n,0,\ldots)},$$

i.e., $s_{-\lambda}$ *is well-defined in* $S(\hat{\underline{\mathfrak{h}}}_{\mathbf{Z}}^{-})$.

Proof: This follows from (4.3.4) and the fact that for $w \in S_n$, $w\delta - \delta$ may be expressed as $(1 - w^{-1}(1), \ldots, n+1 - w^{-1}(n+1))$ (cf. [L-P]). □

Let $\lambda \in \mathbf{Z}^n$ and let z_i, $i \in \mathbf{N}$, be indeterminates. Set

$$z^\lambda = z_1^{\lambda_1} z_2^{\lambda_2} \cdots z_n^{\lambda_n} \in \mathbf{C}[z_1^{\pm 1}, \ldots, z_n^{\pm 1}]. \tag{4.3.9}$$

For $w \in S_n$, set

$$wz^\lambda = z^{w\lambda}. \tag{4.3.10}$$

Next define the polynomial $\Delta_n(z)$ by

$$\Delta_n(z) = \prod_{i<j}(z_i - z_j) = \sum_{w \in S_n} \varepsilon(w) z^{w\delta_n}. \tag{4.3.11}$$

Lemma 4.3.4 *Let $w \in S_n$. Then we have $w(\Delta_n(z)) = \varepsilon(w)\Delta_n(z)$.*

Proof: The identity (4.3.13) together with the fact that $\varepsilon(w')\varepsilon(w) = \varepsilon(w'w)$ give

$$\begin{aligned} w(\Delta_n(z)) &= w\left(\sum_{w' \in S_n} \varepsilon(w') z^{w'\delta_n}\right) \\ &= \sum_{w' \in S_n} \varepsilon(w') z^{ww'\delta_n} \\ &= \varepsilon(w) \sum_{w' \in S_n} \varepsilon(w)\varepsilon(w') z^{ww'\delta_n} \\ &= \varepsilon(w) \sum_{ww' \in S_n} \varepsilon(ww') z^{ww'\delta_n} \\ &= \varepsilon(w)\Delta_n(z). \ \square \end{aligned}$$

Proposition 4.3.5 *Let $n > 0$. Then we have the following:*

$$E^-(-\alpha, z_1) \cdots E^-(-\alpha, z_n) = \sum_{\lambda \in \mathbf{Z}^n} h_{-\lambda} z^\lambda \tag{4.3.12}$$

$$\Delta_n(z_1, \ldots, z_n) E^-(-\alpha, z_1) \cdots E^-(-\alpha, z_n) = \sum_{\lambda \in \mathbf{Z}^n} s_{-\lambda} z^{(\lambda + \delta_n)}. \tag{4.3.13}$$

Proof: See Proposition 7.2 of [L-P]. □

Proposition 4.3.6 *The family*

$$\{s_{-\lambda} \mid \lambda \textit{ is a partition}\} \tag{4.3.14}$$

is a basis of $S(\hat{\underline{\mathfrak{h}}}_{\mathbf{Z}}^-)$. Moreover, the $\mathbf{Z}$-subalgebra generated by (4.3.14) is a $\mathbf{Z}$-form of $S(\hat{\underline{\mathfrak{h}}}_{\mathbf{Z}}^-)$.

Proof: The first statement is Proposition 7.3(a) in [L-P].

Since the $\mathbf{Z}$-subalgebra generated by the family (4.3.6) is a $\mathbf{Z}$-form for $S(\hat{\underline{h}}_{\mathbf{Z}}^{-})$, then (4.3.8) and (4.3.13) imply the same is true for the $\mathbf{Z}$-subalgebra generated by the family (4.3.14) (cf. [Mac]). □

We next assume that $L = \sum_{i=1}^{n} \mathbf{Z}\alpha_i$ is an even nondegenerate lattice. For this choice of L we have

$$\begin{aligned} S(\hat{\underline{h}}_{\mathbf{Z}}^{-}) &= \mathbf{C}[\alpha_1(-1), \alpha_1(-2), \ldots, \alpha_2(-1), \ldots, \alpha_n(-1), \ldots] \\ &= \mathbf{C}[\alpha_1(-1), \ldots][\alpha_2(-1), \ldots] \cdots [\alpha_n(-1), \ldots]. \end{aligned}$$

Let $s_{i,-\lambda_i}$ denote the Schur polynomial $s_{-\lambda_i}$ in the variables $\alpha_i(-n)$. Thus as a result of Proposition 4.3.6 we have the following description of a basis for $S(\hat{\underline{h}}_{\mathbf{Z}}^{-})$.

Proposition 4.3.7 *The set*

$$S_L = \{\prod_{i=1}^{n} s_{i,-\lambda_i} \mid \lambda_i \textit{ is a partition}\}$$

is a basis for $S(\hat{\underline{h}}_{\mathbf{Z}}^{-})$. *Moreover, the* $\mathbf{Z}$*-subalgebra generated by* S_L *is a* $\mathbf{Z}$*-form of* $S(\hat{\underline{h}}_{\mathbf{Z}}^{-})$.

4.4 An integral form for the vertex algebra $V_{L'}$

In this section we weaken the conditions on the lattice L (cf. §4.2) by allowing L to be possibly non-positive definite. Then we form the vertex algebra V_L (as opposed to the vertex operator algebra of the earlier sections). The presentation which follows here, as in the next two sections, will be a more detailed version of the approach Borcherds uses in his announcement [B]. Again note that since vertex operator algebras are vertex algebras, all of the results here also hold for V_L with L positive definite.

Let L be an even nondegenerate lattice. We *do not* assume that L is positive definite. Form V_L and the vertex operators $Y(\cdot, z)$ as in §4.1. Notice that conditions (4.1.2) and (4.1.3) can not be guaranteed.

In §4.2, we defined a bilinear form $\langle\cdot,\cdot\rangle$ on

$$(V_L)_{(1)} \oplus \mathbf{C}\iota(1)_{-1} \oplus \mathbf{C}(-L(0) + \tfrac{1}{24}dim\,\underline{h}\,\iota(1)_{-1}).$$

Since we will eventually embed the derived subalgebra $\underline{\ell}'$ in a quotient module of V_L, we wish to also have a symmetric bilinear form on $V_L \times V_L$. First we define $(\cdot,\cdot)$ on $\mathbf{C}\{L\} \times \mathbf{C}\{L\}$ by

$$(\iota(a), \iota(b)) = \delta_{a,b}. \tag{4.4.1}$$

Then extend $(\cdot,\cdot)$ to $V_L \times V_L$ so that for a, $b \in \hat{L}$ and $h \in \underline{h}$, we have

$$(h(-m)\cdot\iota(a), \iota(b)) = (\iota(a), h(m)\cdot\iota(b)) \tag{4.4.2}$$

and in general,

$$(h(-m)\cdot v, w) = (v, h(m)\cdot w) \tag{4.4.3}$$

holds true for $h \in \underline{h}$ and v, $w \in V_L$. The form $(\cdot,\cdot)$ defined above is the unique symmetric bilinear form (cf. [B], [F-L-M]) on V_L which is contravariant.

Recall from (4.1.20) that the $\hat{L}$-induced module $\mathbf{C}\{L\}$ is isomorphic to the quotient of the group algebra $\mathbf{C}[\hat{L}]$ by the two-sided ideal $(1+\kappa)\mathbf{C}[\hat{L}]$. Thus $\mathbf{C}\{L\}$ has an algebra structure given by $\iota(a)\iota(b) = \iota(ab)$ where a, $b \in \hat{L}$. Since $V_L = S(\hat{\underline{h}}_{\mathbf{Z}}^-) \otimes \mathbf{C}\{L\}$, V_L may also be viewed as an algebra.

In the previous section we were able to describe an integral form for the $\tilde{\underline{h}}$-module $S(\hat{\underline{h}}_{\mathbf{Z}}^-)$. We now want to give a description of a $\mathbf{Z}$-form for the vertex algebra V_L.

Define V' to be the $\mathbf{Z}$-subring of V_L spanned by

$$\{f \otimes \iota(a) \mid f \in \mathbf{S}_L,\ a \in \hat{L}\}. \tag{4.4.4}$$

From Proposition 4.3.7 we know that the $\mathbf{Z}$-span of the set $\mathbf{S}_L$ is an integral form of $S(\hat{\underline{h}}_{\mathbf{Z}}^-)$. We next observe that $V' \otimes_{\mathbf{Z}} \mathbf{C} = V_L$ since $\mathbf{S}_L \otimes_{\mathbf{Z}} \mathbf{C} = S(\hat{\underline{h}}_{\mathbf{Z}}^-)$ and $\{\iota(a) \mid a \in \hat{L}\}$ is a basis of $\mathbf{C}\{L\}$. Let $\overline{V'}$ denote the closure of V' with respect to the action of the operators $\frac{(L(-1))^m}{m!}$ for $m \in \mathbf{N}$. Note that since $V' \subseteq \overline{V'}$, we have $\overline{V'}$ is also an integral form for V_L.

Theorem 4.4.1 *Let $(V_L)_{\mathbf{Z}}$ be the smallest $\mathbf{Z}$-subring of V_L containing all the $\iota(a)$ for $a \in L$ and closed under action by the operators $\frac{(L(-1))^m}{m!}$, $m \in \mathbf{N}$. Then $(V_L)_{\mathbf{Z}} = \overline{V'}$, i.e., $(V_L)_{\mathbf{Z}}$ is a $\mathbf{Z}$-form of V_L.*

Proof: Let V'' be any $\mathbf{Z}$-subring of V_L such that V'' contains $\iota(a)$ for all $a \in \hat{L}$ and is closed under action by $\frac{(L(-1))^m}{m!}$. So for $a \in \hat{L}$ with $\bar{a} = \alpha_i$ and $i = 1, 2, \ldots, n$, formula (8.8.5) of [F-L-M] (which is a consequence of (4.1.13)) implies that

$$e^{L(-1)z} \cdot \iota(a) = \exp\left(\textstyle\sum_{n \geq 1} \frac{\alpha_i(-n)}{n} z^n\right)\iota(a) = E^-(-\alpha_i, z)\iota(a). \tag{4.4.5}$$

We have $E^-(-\alpha_i, z)\iota(a) = \left(\sum_{\lambda \in \mathbf{Z}} s_{i,-\lambda} z^\lambda\right) \otimes \iota(a)$ by Proposition 4.3.5. Thus multiplying both sides of (4.4.5) on the right by $\iota(a^{-1})$ and then equating the coefficients of like powers of z in (4.4.5), we see that $s_{-\lambda} \otimes \iota(1) = h_{-\lambda} \otimes \iota(1) \in V''$ for $\lambda \in \mathbf{Z}$. By definition V'' is a $\mathbf{Z}$-subring of V_L; thus for any two μ, $\lambda \in \mathbf{Z}$, the product $h_{-\mu}h_{-\lambda} \otimes \iota(1) \in V''$. By Proposition 4.3.5 we observe that

$$\begin{aligned}\left(\sum_{\lambda \in \mathbf{Z}} h_{-\lambda} z_1^\lambda \otimes \iota(1)\right)\left(\sum_{\mu \in \mathbf{Z}} h_{-\mu} z_2^\mu \otimes \iota(1)\right) &= E^-(-\alpha_i, z_1)E^-(-\alpha_i, z_2) \otimes \iota(1) \\ &= \sum_{\lambda \in \mathbf{Z}^2} h_{-\lambda} z^\lambda \otimes \iota(1),\end{aligned}$$

and in general, we see that

$$\prod_{j=1}^n \left(\sum_{\lambda_j \in \mathbf{Z}} h_{-\lambda_j} z_j^{\lambda_j} \otimes \iota(1)\right) = \prod_{j=1}^n E^-(-\alpha_i, z_j) \otimes \iota(1) = \sum_{\lambda \in \mathbf{Z}^n} h_{-\lambda} z^\lambda \otimes \iota(1). \tag{4.4.6}$$

Thus $h_{-\lambda} \otimes \iota(1) \in V''$ for $\lambda \in \mathbf{Z}^n$. By (4.3.8) we also have $s_{-\lambda} \otimes \iota(1) \in V''$ for $\lambda \in \mathbf{Z}^n$. Hence V'' contains the set (4.4.4), and so $V' \subset V''$. The closure $\overline{V'}$ of V' with respect to the action of the operators $\frac{(L(-1))^m}{m!}$ for $m \in \mathbf{N}$ is also contained in each $\mathbf{Z}$-subring V''. Thus $\overline{V'}$ is the smallest $\mathbf{Z}$-subring of V_L containing all the $\iota(a)$ for $a \in \hat{L}$ and closed under the action of $\frac{(L(-1))^m}{m!}$. Hence $\overline{V'} = (V_L)_{\mathbf{Z}}$ is a $\mathbf{Z}$-form for V_L. □

Proposition 4.4.2 *If u and v are elements in $(V_L)_{\mathbf{Z}}$, then for $n \in \mathbf{Z}$, $u_n \cdot v$ is also in $(V_L)_{\mathbf{Z}}$.*

Proof: Recall that $u_n \cdot v$ is given by the coefficient of z^{-n-1} in the expression $Y(u,z) \cdot v$. First assume that

$$u_n \cdot v \in (V_L)_{\mathbf{Z}} \tag{4.4.7}$$

for all u, $v \in V'$. Since $(V_L)_{\mathbf{Z}} = \overline{V'}$ we may assume without loss of generality that

$$u = \frac{L(-1)^k}{k!} \cdot u' \text{ and } v = \frac{L(-1)^l}{l!} \cdot v'$$

with u', $v' \in V'$. Then by (4.1.13) we have

$$Y\left(\frac{(L(-1))^k}{k!} \cdot u', z\right) \cdot v = \frac{1}{k!}\left(\frac{d}{dz}\right)^k Y(u', z) \cdot v. \tag{4.4.8}$$

Since $\frac{1}{k!}\left(\frac{d}{dz}\right)^k$ preserves $(V_L)_{\mathbf{Z}}\,[[z, z^{-1}]]$, (4.4.8) shows that

$$\text{if } u'_n \cdot v \in (V_L)_{\mathbf{Z}} \text{ for all } n \in \mathbf{Z}, \text{ then } u_n \cdot v \in (V_L)_{\mathbf{Z}} \text{ for all } n. \tag{4.4.9}$$

Also Proposition 8.8.3 in [F-L-M] gives

$$Y(u,z) \cdot v = e^{zL(-1)}(Y(v,-z) \cdot u). \tag{4.4.10}$$

So if we equate the coefficients of z^{-n-1} on both sides of (4.4.10) we obtain

$$u_n \cdot v = \sum_{i \geq 0} (-1)^{-n-1-i} \frac{(L(-1))^i}{i!}(v_{n+i} \cdot u), \tag{4.4.11}$$

(cf. [B]) and, replacing u by u' in (4.4.11), we have

$$u'_n \cdot v = \sum_{i \geq 0} (-1)^{-n-1-i} \frac{(L(-1))^i}{i!}(v_{n+i} \cdot u'). \tag{4.4.12}$$

Thus by our assumption (4.4.7) we have $(v')_{n+i} \cdot u' \in (V_L)_{\mathbf{Z}}$ for all $n+i \in \mathbf{Z}$. Then (4.4.12) shows that $u'_n \cdot v \in (V_L)_{\mathbf{Z}}$ for all $n \in \mathbf{Z}$. Applying (4.4.9) again gives $u_n \cdot v \in (V_L)_{\mathbf{Z}}$ for all $n \in \mathbf{Z}$. Thus it remains to establish (4.4.7).

Let $a_1, a_2, \ldots, a_k, b_1, b_2, \ldots, b_l \in \hat{L}$ be such that

$$a = a_1 \cdots a_k \text{ and } b = b_1 \cdots b_l. \tag{4.4.13}$$

Set

$$\begin{aligned} A &= \exp\left(\textstyle\sum_{i=1}^k \sum_{n \geq 1} \frac{\bar{a}_i(-n)}{n} w_i^n\right) \cdot \iota(a) \text{ and} \\ B &= \exp\left(\textstyle\sum_{j=1}^l \sum_{n \geq 1} \frac{\bar{b}_j(-n)}{n} x_j^n\right) \cdot \iota(b). \end{aligned} \tag{4.4.14}$$

From [F-L-M] and [Mac] we observe that the coefficients in the formal power series A and B span $S(\hat{\underline{h}}_{\mathbf{Z}}^{-}) \otimes \iota(a)$ and $S(\hat{\underline{h}}_{\mathbf{Z}}^{-}) \otimes \iota(b)$ (cf. Proposition 4.3.1, Proposition 4.3.5), respectively, and we also see that these coefficients are elements of V'. Thus it is enough to show the coefficients of $(A)_n(B)$ are elements in $(V_L)_{\mathbf{Z}}$.

We compute the formal power series $(A)_n(B)$ using the methods developed in Chapter 8 of [F-L-M]. We have

$$A = {}^\circ_\circ Y(a_1, w_1) \cdots Y(a_k, w_k){}^\circ_\circ \cdot \iota(1) \tag{4.4.15}$$

since by (4.1.32), (3.1.2), (4.4.13) and (4.4.14),

$$\begin{aligned} A &= \exp\left(\textstyle\sum_{i=1}^k \sum_{n\geq 1} \frac{\overline{a}_i(-1)}{n} w_i^n\right) \cdot \iota(a) \\ &= \prod_{i=1}^k \exp\left(\textstyle\sum_{n\geq 1} \frac{\overline{a}_i(-n)}{n} w_i^n\right) \cdot \iota(a_1 \cdots a_k) \\ &= {}^\circ_\circ Y(a_1, w_1) \cdots Y(a_k, w_k){}^\circ_\circ \iota(1). \end{aligned}$$

Similarly,

$$B = {}^\circ_\circ Y(b_1, x_1) \cdots Y(b_l, x_l){}^\circ_\circ \cdot \iota(1). \tag{4.4.16}$$

Using the fact that

$$e^{y(\frac{d}{dz})} v(z) = v(z+y) = v(z(1+\frac{y}{z})) \tag{4.4.17}$$

for $v(z) \in V[[z, z^{-1}]]$ and $y \in w\mathbf{C}[[w]]$, $y \neq 0$ (cf. [F-L-M]), and (4.1.32) we have

$$Y(A, z) = {}^\circ_\circ Y(a_1, z+w_1) \cdots Y(a_k, z+w_k){}^\circ_\circ. \tag{4.4.18}$$

Thus we see

$$\begin{aligned} Y(A,z)\cdot B &= {}^\circ_\circ Y(a_1, z+w_1) \cdots Y(a_k, z+w_k){}^\circ_\circ \cdot {}^\circ_\circ Y(b_1, x_1) \cdots Y(b_l, x_l){}^\circ_\circ \cdot \iota(1) \\ &= E^-(-\overline{a}_1, z+w_1) \cdots E^-(-\overline{a}_k, z+w_k) E^+(-\overline{a}_1, z+w_1) \cdots E^+(-\overline{a}_k, z+w_k) \\ &\quad \cdot \iota(a)(z+w_1)^{\overline{a}_1} \cdots (z+w_k)^{\overline{a}_k} \cdot E^-(-\overline{b}_1, x_1) \cdots E^-(-\overline{b}_l, x_l)\iota(b) \\ &= E^-(-\overline{a}_1, z+w_1) \cdots E^-(-\overline{a}_k, z+w_k) E^-(-\overline{b}_1, x_1) \cdots E^-(-\overline{b}_l, x_l)\iota(ab) \times \\ &\quad \prod_{\substack{1\leq i\leq k \\ 1\leq j\leq l}} z^{\langle \overline{a}_i, \overline{b}_j\rangle} \left(1 - \frac{(w_i - x_j)}{z}\right)^{\langle \overline{a}_i, \overline{b}_j\rangle}. \end{aligned}$$

Since L is even, each $\langle \overline{a}_i, \overline{b}_j \rangle \in \mathbf{Z}$ and so the formal power series

$$\prod_{\substack{1 \le i \le k \\ 1 \le j \le l}} z^{\langle \overline{a}_i, \overline{b}_j \rangle} \left(1 - \frac{(w_i - x_j)}{z}\right)^{\langle \overline{a}_i, \overline{b}_j \rangle}$$

has integer coefficients. Also note that (4.3.11) gives

$$z^{-\delta_n} \Delta_n(z) = \prod_{1 \le i < j \le n} \left(1 - \frac{z_j}{z_i}\right).$$

Thus Proposition 4.3.5 and the above imply $\sum_\lambda h_{-\lambda} z^\lambda = \prod_{1 \le i < j \le n} \left(1 - \frac{z_j}{z_i}\right)^{-1} \cdot \sum_\lambda s_{-\lambda} z^\lambda$. Hence by Proposition 4.3.5, (4.4.19) and Theorem 4.4.1, $Y(A, z) \cdot B$ is an element of $(V_L)_{\mathbf{Z}}\,[[z^{\pm 1}, w_1^{\pm 1}, \ldots, w_k^{\pm 1}, x_1^{\pm 1}, \ldots, x_l^{\pm 1}]]$. □

4.5 The embedding of the derived subalgebra $\underline{\ell}'$ in a module for $V_{L'}$

Let L be the root lattice of the simply-laced affine algebra $\underline{\ell}$, i.e., $L = \oplus_{i=0}^n \mathbf{Z}\alpha_i$. Set $L' = \mathbf{Z}\beta \oplus L$ where $\beta \notin L$ and extend $\langle \cdot, \cdot \rangle$ to $L' \times L'$ by

$$\begin{aligned} \langle \cdot, \cdot \rangle|_{L \times L} &= \langle \cdot, \cdot \rangle \\ \langle \alpha_i, \beta \rangle = \langle \beta, \alpha_i \rangle &= 0 \;\text{ for } 1 \le i \le n \\ \langle \alpha_0, \beta \rangle = \langle \beta, \alpha_0 \rangle &= -1 \\ \langle \beta, \beta \rangle &= 2. \end{aligned} \tag{4.5.1}$$

Thus L' is an even nondegenerate lattice that is not positive definite. For instance,

$$\langle a_0\alpha_0 + a_1\alpha_1 + \cdots + a_n\alpha_n + \beta, a_0\alpha_0 + a_1\alpha_1 + \cdots + a_n\alpha_n + \beta \rangle = 0$$

where $a_0\alpha_0 + \cdots + a_n\alpha_n = \delta$, the unique nonzero vector such that $\langle \delta, \alpha_i \rangle = 0$ ([Kac] or [K-K-L-W]). Note that since L' is not positive definitive, $V_{L'}$ is a vertex algebra which is not a vertex operator algebra.

A *module* W for the vertex algebra $V_{L'}$ is a $\mathbf{Q}$-graded vector space

$$W = \coprod_{n \in \mathbf{Q}} W_{(n)} \text{ for } w \in W_{(n)};\; n = \mathrm{wt}\, w \tag{4.5.2}$$

equipped with a linear map

$$\begin{aligned} V_{L'} &\longrightarrow (End\, W)[[z, z^{-1}]] \\ v &\longmapsto Y(v,z) = \textstyle\sum_{n\in\mathbf{Z}} v_n z^{-n-1} \text{ (with } v_n \in End\, W) \end{aligned} \tag{4.5.3}$$

satisfying:

$$v_n \cdot w = 0 \text{ for } n \text{ sufficiently large;} \tag{4.5.4}$$

$$Y(1,z) = 1; \tag{4.5.5}$$

$$\begin{aligned} z_0^{-1}\delta\left(\frac{z_1 - z_2}{z_0}\right) Y(u,z_1)Y(v,z_2) - z_0^{-1}\delta\left(\frac{z_1 - z_2}{-z_0}\right) Y(v,z_2)Y(u,z_1) \\ = z_2^{-1}\delta\left(\tfrac{z_1 - z_0}{z_2}\right) Y(Y(u,z_0)v, z_2); \end{aligned} \tag{4.5.6}$$

$$\begin{aligned} [L(m), L(n)] = (m-n)L(m+n) + \tfrac{1}{12}(m^3 - m)\delta_{m+n,\,0}(\text{rank } V_{L'}) \\ \text{for } m,\ n \in \mathbf{Z}, \text{ and } L(m),\ L(n) \text{ as in (4.1.10);} \end{aligned} \tag{4.5.7}$$

$$L(0)\cdot w = nw = (\text{wt } w)w \text{ for } n \in \mathbf{Q} \text{ and } w \in W_{(n)}; \tag{4.5.8}$$

$$\tfrac{d}{dz} Y(u,z) = Y(L(-1)\cdot u, z); \tag{4.5.9}$$

$$[L(-1), Y(v,z)] = Y(L(-1)\cdot v, z) \tag{4.5.10}$$

$$[L(0), Y(v,z)] = Y(L(0)\cdot v, z) + zY(L(-1)\cdot v, z) \tag{4.5.11}$$

for $u,\ v \in V_{L'}$ and $w \in W$. Note that in (4.5.6) $Y(u,z_0)\cdot v$ is an operator on $V_{L'}$ and in (4.5.9) - (4.5.11), $L(-1)$ is acting on $V_{L'}$ (cf. [F-L-M], [B]).

Set $DV_{L'}$ equal to the sum over all $i \geq 1$ of the spaces $\frac{(L(-1))^i}{i!}(V_{L'})$. Similarly, for any subspaces W of $V_{L'}$, set DW equal to the sum over all $i \geq 1$ of the subspaces $\frac{(L(-1))^i}{i!}(W)$.

Proposition 4.5.1 *$V_{L'}/DV_{L'}$ with product defined by*

$$[\overline{u}, \overline{v}] = u_0 \cdot v + DV_{L'} \tag{4.5.12}$$

for $u,\ v \in V_{L'}$, $\overline{u} = u + DV_{L'}$ and $\overline{v} = v + DV_{L'}$, is a Lie algebra.

Proof: If $u = L(-1)\cdot u'$ for some $u' \in V_{L'}$, then by (4.1.13) $u_0 = 0$. So if $u \in DV_{L'}$, $v \in V_{L'}$, then $u_0 \cdot v = 0$, Thus $u_0 \cdot v \in DV_{L'}$. Furthermore for any $u \in V_{L'}$ and any

$v \in DV_{L'}$, (4.4.11) and (4.1.13) imply that $u_0 \cdot v \in DV_{L'}$. Thus $[\overline{u}, \overline{v}]$ is well-defined on $V_{L'}/DV_{L'}$.

Next let u and v be in $V_{L'}$. Again using (4.4.11), we obtain

$$[\overline{u}, \overline{v}] = -v_0 \cdot u + DV_{L'} = -[\overline{v}, \overline{u}].$$

For the Jacobi identity for Lie algebras, we use a component version of the Jacobi identity for vertex algebras (4.1.8). Since as formal power series,

$$z_2^{-1}\delta\left(\frac{z_1 - z_0}{z_2}\right) = z_1^{-1}\delta\left(\frac{z_2 + z_0}{z_1}\right) \tag{4.5.13}$$

(cf. [F-L-M]), we may use (4.5.13) to rewrite (4.1.8) as

$$\begin{aligned} z_0^{-1}\sum_{k\in\mathbf{Z}}(z_1 - z_2)^k z_0^{-k} Y(u, z_1)Y(v, z_2) - z_0^{-1}\sum_{k\in\mathbf{Z}}(z_2 - z_1)^k(-z_0)^{-k} Y(v, z_2)Y(u, z_1) \\ = z_1^{-1}\textstyle\sum_{k\in\mathbf{Z}}(z_2 + z_0)^k z_1^{-k} Y(Y(u, z_0)v, z_2). \end{aligned} \tag{4.5.14}$$

Now apply both sides of (4.5.14) to a vector $w \in V_{L'}$. Equating the coefficient of $z_0^{-m-1}z_2^{-n-1}z_1^{-1}$ on both sides, we obtain

$$(u_m \cdot v)_n \cdot w = \sum_{i\geq 0}(-1)^i \binom{m}{i} \left(u_{m-i} \cdot (v_{n+i} \cdot w) - (-1)^m v_{m+n-i} \cdot (u_i \cdot w)\right) \tag{4.5.15}$$

(cf. [B]). Thus for u, v and $w \in V_{L'}$, (4.5.15) gives

$$[[\overline{u}, \overline{v}], \overline{w}] = u_0 \cdot (v_0 \cdot w) - v_0 \cdot (u_0 \cdot w) + DV_{L'} = [\overline{u}, [\overline{v}, \overline{w}]] - [\overline{v}, [\overline{u}, \overline{w}]]. \ \square$$

Corollary 4.5.2 *If W is a module for $V_{L'}$, then W is a $V_{L'}/DV_{L'}$ module by letting $v \in V_{L'}/DV_{L'}$ act as the operator v_0 on W.*

Proof: Apply (4.5.15). □

We would like to realize the derived subalgebra of $\underline{\ell}$ as a subalgebra of the Lie algebra $V_{L'}/DV_{L'}$. Once this is achieved, any module for the vertex algebra $V_{L'}$ is automatically a module for the algebra $\underline{\ell}$ by Corollary 4.5.2. Since we have described an integral form of $V_{L'}$, we can also examine how the $\mathbf{Z}$-forms of $\underline{\ell}$ and $U(\underline{\ell})$ act on $(V_{L'})_{\mathbf{Z}}$. While the quotient $V_{L'}/DV_{L'}$ is a Lie algebra, it is generally much larger than

$\underline{\ell}'$. We now set out to describe a smaller subalgebra of $V_{L'}/DV_{L'}$ which is "closer" in size to $\underline{\ell}'$.

Define P^i to be the subspace of $V_{L'}$ [B] given by

$$P^i = \{v \in V_{L'} \mid L(n) \cdot v = iv \text{ if } n = 0 \text{ and } 0 \text{ if } n \geq 1\}$$

(cf. (8.7.34) and (8.7.35) in [F-L-M]).

Proposition 4.5.3 *If $v \in P^1$, then the operator v_0 commutes with $L(n)$ for $n \in \mathbf{Z}$. Hence v_0 preserves all the spaces P^i.*

Proof: To see that v_0 preserves each space P^i, we compute the commutator of the operators v_0 and $L(n)$ for $n \geq 0$. Recall that the residue in z_0^{-1} of (4.1.8) gives the bracket formula (4.1.33). Using (4.1.33) and (4.1.13) with $u = \omega$ and $v \in P^1$, we obtain

$$[L(z_1), Y(v, z_2)] = \mathrm{Res}_{z_0} z_2^{-1} Y(L(z_0)v, z_2) e^{-z_0 \frac{\partial}{\partial z_1}} \delta(z_1/z_2). \tag{4.5.16}$$

The right-hand side of (4.5.16) reduces to

$$z_2^{-1} \frac{d}{dz_2} Y(v, z_2) \delta(z_1/z_2) - z_2^{-1} Y(v, z_2) \frac{\partial}{\partial z_1} \delta(z_1/z_2). \tag{4.5.17}$$

Then equating the coefficients of $z_1^{-n-2} z_2^{-1}$ on both sides of (4.5.17), we have

$$[L(n), v_0] = (-n-1)v_n + (n+1)v_n = 0. \tag{4.5.18}$$

Hence (4.5.18) shows that v_0 preserves all the eigenspaces for $L(n)$ where $n \geq 0$ (cf. Proposition 8.7.7 in [F-L-M]). □

Proposition 4.5.4 *If $v = L(-1) \cdot u$ for $v \in P^1$ and $u \in V_{L'}$, then there exists an element $u' \in P^0$ such that $v = L(-1)u'$.*

Proof: Let $v \in P^1$ with $v = L(-1) \cdot u$ for $u \in V_{L'}$. By (4.1.9) we see that $L(0) \cdot v = L(-1) \cdot (L(0) \cdot u) + v$. Hence $L(-1) \cdot (L(0) \cdot u) = 0$. We may write u as a (finite) sum of its homogeneous parts, i.e.,

$$u = \sum_i u^{(i)} \text{ where } u^{(i)} \in (V_{L'})_{(i)}. \tag{4.5.19}$$

Thus we have $0 = L(-1)\cdot\left(L(0)\cdot\sum_i u^{(i)}\right) = \sum_i iL(-1)u^{(i)}$. So $L(-1)\cdot u^{(i)}$ must equal zero for all $i \neq 0$. Next put $u' = u - \sum_{i\neq 0} u^{(i)}$. Then we see that

$$L(-1)\cdot u' = L(-1)\cdot u = v \text{ and } L(0)\cdot u' = 0.$$

It remains to see that $L(n)\cdot u' = 0$ for $n > 0$. By (4.1.5) there exists an $n \in \mathbb{Z}$ so that $L(n)\cdot u' = 0$ for all $m \geq n$. Using (4.1.9) again, we observe that

$$L(n-1)\cdot u' = \frac{1}{n+1}L(n)\cdot(L(-1)\cdot u') - \frac{1}{n+1}L(-1)\cdot(L(n)\cdot u') = 0$$

since $L(-1)\cdot u' = v$ and $L(n)\cdot u' = 0$. Thus an induction argument allows us to conclude that $L(m)\cdot u' = 0$ for $m > 0$. Hence $u' \in P^0$. □

From Proposition 4.5.4 we see that $P^1 \cap DV_{L'} \subset DP^0$. To obtain the reverse inclusion, we notice that if $v \in P^0$, then

$$L(n)\cdot(L(-1)\cdot v) = \begin{cases} L(-1)\cdot v & \text{if } n = 0 \\ 0 & \text{if } n \geq 1. \end{cases}$$

Thus we have $P^1 \cap DV_{L'} = DP^0$. Hence P^1/DP^0 is a Lie subalgebra of $V_{L'}/DV_{L'}$. □

Proposition 4.5.5 ([B]) *Let $\underline{\ell}$ be an affine Lie algebra constructed in §4.2. Choose a section $e : L' \to \hat{L}'$ such that condition (2.2.10) holds. The derived subalgebra $\underline{\ell}'$ can be mapped to the Lie algebra P^1/DP^0 by*

$$\begin{aligned} e_i &\longmapsto \iota(e_{\alpha_i}) \\ f_i &\longmapsto \epsilon(\alpha_i, -\alpha_i)\iota(e_{-\alpha_i}) \\ h_i &\longmapsto \alpha_i(-1)\otimes\iota(1) \end{aligned} \tag{4.5.20}$$

where α_i are the simple real roots of $\underline{\ell}$ and the e_i, f_i and h_i are the usual Kac-Moody generators. Moreover, this representation of $\underline{\ell}'$ is faithful.

Proof: It is clear by previous definitions that each of the generators $\alpha_i(-1)\otimes\iota(1)$, $\iota(e_{\alpha_i})$ and $\iota(e_{-\alpha_i})$ are elements of P^1. Using Proposition 4.5.1 we see that the Kac-Moody relations for this set of generators are satisfied. The representation of $\underline{\ell}'$ on P^1/DP^0 is faithful since L' is nondegenerate. □

Remark 4.5.6 The choice of the lattice L' allows us to embed $\underline{\ell}'$ into P^1/DP^0. If we had used the positive definite lattice L of the earlier chapters, we would only be able to map the finite dimensional Lie algebra $\underline{g}$ into P^1/DP^0 (cf. [B]).

4.6 A family of irreducible integrable modules for $\underline{\ell}$

Let L'' be the weight lattice associated to the lattice L' of §4.5. (So for $\gamma \in L''$ and for $\alpha \in L'$, $\langle \alpha, \gamma \rangle \in \mathbf{Z}$.) For each element $\alpha \in L''$, we wish to construct a module, denoted by $M(\alpha)$, which is irreducible with highest weight α and is integrable. These modules were first studied in [B].

Remark 4.6.1 We note that besides the $\mathbf{Z}$-grading by its weights, $V_{L'}$ also has an L'-grading by defining

$$\deg \alpha_1(-n_1)\alpha_2(-n_2)\cdots\alpha_k(-n_k) \otimes \iota(a) = \overline{a}$$

where $\alpha_i \in L'$, $a \in \hat{L}'$ and $n_i \in \mathbf{Z}_+$.

We next define the module $M(\alpha)$ for $\alpha \in L'$. Choose a section $e : L' \to \hat{L}'$ satisfying (2.2.10). Take the subspace P^i with $i = \frac{1}{2}\langle \alpha, \alpha \rangle$. Using the L'-grading mentioned in Remark 4.6.1, we see there is a maximal $\underline{\ell}$-submodule M of P^i such that $\iota(e_\alpha) \notin M$. Note that since L' is nondegenerate, all submodules of P^i are L'-graded. So in particular, M is a graded submodule of P^i. Let $M(\alpha)$ be the $\underline{\ell}$-submodule P^i/M.

The following theorem is a collection of properties associated with the $\underline{\ell}$-modules $M(\alpha)$.

Theorem 4.6.2 ([B]) *(1)* $dim\,(M(\alpha))_\alpha = 1$.

(2) *If* $\langle \gamma, \gamma \rangle > \langle \alpha, \alpha \rangle$, *then* $dim\,(M(\alpha))_\gamma = 0$.

(3) $M(\alpha)$ *is integrable.*

(4) $dim\,(M(\alpha))_\gamma < \infty$ *for each* $\gamma \in L''$.

(5) $M(\alpha)$ *is irreducible.*

(6) *If* γ *and* γ' *are conjugate under the Weyl group, then* $M(\gamma) = M(\gamma')$.

(7) *For α a highest (or lowest) weight, the module $M(\alpha)$ is the corresponding standard weight module with highest weight vector $\iota(e_\alpha)$ (lowest weight vector $\iota(e_{-\alpha})$). If α is a real root of $\underline{\ell}'$, then $M(\alpha)$ is a quotient of the adjoint representation.*

Proof: For (1) and (2) we see that $\iota(e_\alpha) \in P^{\frac{1}{2}\langle\alpha,\alpha\rangle}$, so $dim\ (M(\alpha))_\alpha \geq 1$. Recall that

$$L(0)\cdot(\gamma_1(-n_1)\cdots\gamma_r(-n_r)\otimes\iota(e_\tau)) = n_1+\cdots+n_r+\frac{1}{2}\langle\tau,\tau\rangle.$$

Thus if $\gamma_1(-n_1)\cdots\gamma_r(-n_r)\otimes\iota(e_\tau)\in P^{\frac{1}{2}\langle\alpha,\alpha\rangle}$, we have

$$n_1+\cdots+n_r = \frac{1}{2}\left(\langle\alpha,\alpha\rangle - \langle\tau,\tau\rangle\right).$$

So if $\langle\alpha,\alpha\rangle \leq \langle\tau,\tau\rangle$ and all $n_i > 0$, we must have $r=0$ and $\langle\alpha,\alpha\rangle = \langle\tau,\tau\rangle$. The result then follows.

For property (3) we let $v = f\otimes\iota(e_\gamma)$, where $f\in S(\hat{\underline{h}}_{\mathbf{Z}}^-)$ and $\gamma\in L'$. Note that any element of $M(\alpha)$ is a linear combination of such v. For any real root τ of $\underline{\ell}$, we observe that

$$\iota(e_\tau)^n\cdot(f\otimes\iota(e_\gamma)) = \epsilon(\tau,\frac{n(n-1)}{2}\tau+n\gamma)f\otimes\iota(e_{n\tau+\gamma}).$$

Thus $\iota(e_\tau)^n\cdot v\in (M(\alpha))_{n\tau+\gamma}$. Now choose $n\in\mathbf{Z}_+$ so that

$$\langle n\tau+\gamma, n\tau+\gamma\rangle = 2n^2+2n\langle\gamma,\tau\rangle+\langle\gamma,\gamma\rangle > \langle\alpha,\alpha\rangle.$$

Then by (2) $\iota(e_\tau)^n\cdot v = 0$. Hence $M(\alpha)$ is integrable.

To see (4) holds, we notice that since the $L(0)$-weight spaces of $S(\hat{\underline{h}}_{\mathbf{Z}}^-)$ are finite dimensional, we must have $dim\ (M(\alpha))_\gamma < \infty$ for any weight γ.

Since P^1/M is integrable, it is also completely reducible ([Kac], Theorem 10.7). So any decomposition of P^1/M into a direct sum of its irreducibles must contain only one nonzero summand by the maximality of M. Hence $M(\alpha)$ is irreducible.

Suppose that γ and γ' are conjugate under the Weyl group. Since $M(\gamma)$ and $M(\gamma')$ are integrable, Proposition 3.7 of [Kac] and (1) imply that

$$dim\ (M(\gamma))_\gamma = dim\ (M(\gamma))_{\gamma'},\ \text{and}\ dim\ (M(\gamma'))_\gamma = dim\ (M(\gamma'))_{\gamma'}.$$

By (1) all of the above dimensions are equal to one. Recall we have

$$M(\gamma) \subseteq P^i/M \text{ and } M(\gamma') \subseteq P^i/M'$$

where $i = \frac{1}{2}\langle\gamma,\gamma\rangle = \frac{1}{2}\langle\gamma',\gamma'\rangle$ and M (M', respectively) is a maximal submodule of P^i not containing $\iota(e_\gamma)$ ($\iota(e_{\gamma'})$, respectively). Since $dim\,(M(\gamma))_{\gamma'} = 1$, we must have $\iota(e_{\gamma'}) \notin M$. Thus $M' \subseteq M$. Similarly we have $\iota(e_\gamma) \notin M'$ and so $M \subseteq M'$. Hence $M = M'$. Since $\overline{\iota(e_\gamma)} \in M(\gamma')$ and $\overline{\iota(e_{\gamma'})} \in M(\gamma)$, property (5) implies $M(\gamma) = M(\gamma')$.

To prove (7), let α be a highest weight. By the maximality of M and (5), $M(\alpha)$ is generated as an $\underline{\ell}$-module by $\overline{\iota(e_\alpha)}$. We know $M(\alpha)$ is integrable and irreducible by (3) and (5), respectively. Thus $M(\alpha)$ is a standard module.

For α a real root of $\underline{\ell}'$, $M(\alpha) \subset P^1$. Hence the action of $\underline{\ell}'$ on $M(\alpha)$ is the adjoint action. □

For $\alpha \in L''$, use the same construction as for $M(\alpha)$, $\alpha \in L'$ with the following exception: replace $V_{L'}$ with $V_{L',\alpha}$ where

$$V_{L',\alpha} = S(\hat{\underline{h}}_{\mathbf{Z}}^{-}) \otimes \mathbf{C}\{L'+\alpha\} = \iota(e_\alpha)\left(S(\hat{\underline{h}}_{\mathbf{Z}}^{-}) \otimes \mathbf{C}\{L'\}\right).$$

Note that for $\alpha \in L'$, $V_{L',\alpha} = V_{L'}$.

As in [B], we define an integral form of $(V_{L',\alpha})_{\mathbf{Z}}$ for $\alpha \in L'' \setminus L'$ to be given by

$$(V_{L',\alpha})_{\mathbf{Z}} = \iota(e_\alpha)\,(V_{L'})_{\mathbf{Z}}\,, \tag{4.6.1}$$

and an integral form of $M(\alpha)$ for $\alpha \in L''$ to be given by

$$M(\alpha)_{\mathbf{Z}} = (V_{L',\alpha})_{\mathbf{Z}} \cap M(\alpha). \tag{4.6.2}$$

Note that if $\alpha \in L'$, then $(V_{L',\alpha}) = V_{L'}$ and $(V_{L',\alpha})_{\mathbf{Z}} = (V_{L'})_{\mathbf{Z}}$. The subspace $\underline{\ell}_{\mathbf{Z}}$ of $\underline{\ell}$, given by the $\mathbf{Z}$-span of the elements of $\underline{\ell}$ represented by the elements of $(V_{L'})_{\mathbf{Z}}$, is a $\mathbf{Z}$-form of $\underline{\ell}$: this is the image of the $\mathbf{Z}$-span of the Chevalley basis elements with the degree operator replaced by an integral multiple of itself (cf. Theorem 2.1.2, Theorem 4.4.1). The new degree operator is given by

$$-\,r(-L(0) + \tfrac{1}{24} dim\,\underline{h}\,\iota(1)_{-1}) \tag{4.6.3}$$

where r is the smallest positive integer such that $r(L(0)\cdot\iota(e_\alpha))\in \mathbf{Z}\alpha$ for all α in L''.

By Proposition 4.4.2 and Proposition 4.5.3 we see that the action of $\underline{\ell}_{\mathbf{Z}}$ preserves $(V_{L',\alpha})_{\mathbf{Z}}$ and $P^{\frac{1}{2}\langle\alpha,\alpha\rangle}$ for $\alpha\in L''$. Thus $\underline{\ell}_{\mathbf{Z}}$ preserves all of the $M(\alpha)_{\mathbf{Z}}$'s.

Let $U^{!}_{\mathbf{Z}}(\underline{\ell})$ denote the $\mathbf{Z}$-subalgebra described in Theorem 4.2.7 with the generator $\Lambda_s(-(-L(0)+\frac{1}{24}dim\,\underline{\mathrm{h}}\,\iota(1)_{-1}))$ replaced by

$$\Lambda_s(-r(-L(0)+\tfrac{1}{24}dim\,\underline{\mathrm{h}}\,\iota(1)_{-1})). \tag{4.6.4}$$

Theorem 4.6.3 *Let $(U(\underline{\ell}))_{\mathbf{Z}}$ denote the subalgebra of $U(\underline{\ell})$ which preserves all the $M(\alpha)_{\mathbf{Z}}$'s for $\alpha\in L''$. Then $U^{!}_{\mathbf{Z}}(\underline{\ell})\subset(U(\underline{\ell}))_{\mathbf{Z}}$.*

Proof: First we show that $(U(\underline{\ell}))_{\mathbf{Z}}$ contains the elements $\frac{(e_i)^n}{n!}$ and $\frac{(f_i)^n}{n!}$ for e_i and f_i as in (4.5.20) and $n\in\mathbf{N}$. To see this is true, it is enough to check that

$$\frac{(e_i)^n}{n!}\cdot(f\otimes\iota(e_\alpha))\in M(\gamma)_{\mathbf{Z}}$$

where $f\in S_L$ (cf. Proposition 4.3.7), α, $\gamma\in L'$, and $f\otimes\iota(e_\alpha)\in M(\gamma)_{\mathbf{Z}}$. By Proposition 4.5.4 we know that

$$\frac{(e_i)^n}{n!}\cdot(f\otimes\iota(e_\alpha))\in M(\gamma).$$

We next show $\frac{(e_i)^n}{n!}$ preserves the integral form of $M(\gamma)$. Recall from Proposition 4.3.5 that f can be represented as a component in the following formal power series expansion:

$$\prod_{j=0}^{k}\{\Delta_{m_j}(w_j)\prod_{s=1}^{m_j}E^-(-\alpha_j,w_{j,s})\}, \tag{4.6.5}$$

where $w_j=(w_{j,1},\ldots,w_{j,m_j})$ and $m_j\in\mathbf{N}$. Thus by Proposition 4.3.5 we have

$$Y(\iota(e_{\alpha_i}),z_1)\cdots Y(\iota(e_{\alpha_i}),z_n)\cdot\prod_{j=0}^{k}\{\Delta_{m_j}(w_j)\prod_{s=1}^{m_j}E^-(-\alpha_j,w_{j,s})\}\otimes\iota(e_\alpha)=$$

$$=\prod_{p=0}^{n-1}\epsilon(\alpha_i,p\alpha_i+\alpha)\cdot\prod_{r=1}^{n}E^-(-\alpha_i,z_r)\cdot\prod_{r<s}\left(1-\frac{z_r}{z_s}\right)^2\cdot z_1^{2(n-1)}\cdots z_{n-1}^2\times$$

$$\prod_{r=1}^{n}E^+(-\alpha_i,z_r)\cdot\prod_{j=0}^{k}\{\Delta_{m_j}(w_j)\prod_{s=1}^{m_j}E^-(-\alpha_j,w_{j,s})\}\otimes\iota(e_{\alpha+n\alpha_i})z^{\langle\alpha,\alpha_i\rangle}\times$$

$$= \prod_{p=0}^{n-1} \epsilon(\alpha_i, p\alpha_i + \alpha) \cdot \prod_{r=1}^{n} E^-(-\alpha_i, z_r) \cdot \Delta_n(z)\Delta_n(z) \cdot \prod_{j=0}^{k} \{\Delta_{m_j}(w_j) \prod_{s=1}^{m_j} E^-(-\alpha_j, w_{j,s})\} \times$$
$$\prod_{j=0} \left\{ \prod_{\substack{1\le r\le n \\ 1\le s\le m_j}} \left(1 - \frac{z_r}{w_{j,s}}\right)^{\langle\alpha_i, \alpha\rangle_j} \right\} \otimes \iota(e_{\alpha+n\alpha_i}) z^{\langle\alpha, \alpha_i\rangle}$$

$$= \prod_{p=0}^{n-1} \epsilon(\alpha_i, p\alpha_i + \alpha) \times$$
$$\sum_{w\in S_n} \varepsilon(w) z^{w\delta_n} \Delta_n(z) \left(\prod_{r=1}^{n} E^-(-\alpha_i, z_r)\right) \left(\prod_{j=0}^{k} \{\Delta_{m_j}(w_j) \prod_{s=1}^{m_j} E^-(-\alpha_j, w_{j,s})\}\right) \times$$
$$\prod_{j=0}^{k} \left\{ \prod_{\substack{1\le r\le n \\ 1\le s\le m_j}} \left(1 - \frac{z_r}{w_{j,s}}\right)^{\langle\alpha_i, \alpha_j\rangle} \right\} \otimes \iota(e_{\alpha+n\alpha_i}) z^{\langle\alpha, \alpha_i\rangle} \times$$

$$= \prod_{p=0}^{n-1} \epsilon(\alpha_i, p\alpha_i + \alpha) \times$$
$$\sum_{w\in S_n} z^{w\delta_n} \Delta_n(wz) \left(\prod_{r=1}^{n} E^-(-\alpha_i, wz_r)\right) \left(\prod_{j=0}^{k} \{\Delta_{m_j}(w_j) \prod_{s=1}^{m_j} E^-(-\alpha_j, w_{j,s})\}\right) \times$$
$$\prod_{j=0}^{k} \left\{ \prod_{\substack{1\le r\le n \\ 1\le s\le m_j}} \left(1 - \frac{wz_r}{w_{j,s}}\right)^{\langle\alpha_i, \alpha_j\rangle} \right\} \otimes \iota(e_{\alpha+n\alpha_i}) (wz)^{\langle\alpha, \alpha_i\rangle}$$

$$= \prod_{p=0}^{n-1} \epsilon(\alpha_i, p\alpha_i + \alpha) \times$$
$$\sum_{w\in S_n} z^{w\delta_n} \left(\sum_{\lambda\in \mathbf{Z}^n} s_{i,-\lambda} z^{(\lambda+\delta_n)}\right) \left(\prod_{j=0}^{k} \sum_{\lambda\in \mathbf{Z}^{m_j}} s_{j,-\lambda} w_j^{(\lambda_{m_j})}\right) \times$$
$$\prod_{j=0} \left\{ \prod_{\substack{1\le r\le n \\ 1\le s\le m_j}} \left(1 - \frac{wz_r}{w_{j,s}}\right)^{\langle\alpha_i, \alpha_j\rangle} \right\} \otimes \iota(e_{\alpha+n\alpha_i}) (wz)^{\langle\alpha, \alpha_i\rangle}.$$

So the coefficient of $(z_1 \cdots z_n)^{-1}$ in the above expression is in the $\mathbf{Z}$-span of $n!(S_{L'}) \otimes \iota(e_{\alpha+n\alpha_i}) \in (V_{L',\alpha})_{\mathbf{Z}}$. Hence $\frac{(e_i)^n}{n!}$ preserves $(V_{L',\alpha})_{\mathbf{Z}}$. An analogous argument works for $\frac{(f_i)^n}{n!}$. The commutation formulas of §3.2 along with the above argument show that $\frac{(e_i)^n}{n!}$, $\frac{(f_i)^n}{n!}$ and $\Lambda_s((e_i)_0 \cdot (f_i))$ are elements of $(U(\underline{\ell}))_{\mathbf{Z}}$ where α is a real root of $\underline{\ell}$ and $s \geq 0$. A straightforward calculation shows $\Lambda_s(-r(-L(0) + \frac{1}{24}dim\, \underline{\mathrm{h}}\, \iota(1)_{-1}))$ is also an element of $(U(\underline{\ell}))_{\mathbf{Z}}$. Hence $U_{\mathbf{Z}}^{\natural}(\underline{\ell}) \subset (U(\underline{\ell}))_{\mathbf{Z}}$. □

We next demonstrate that in fact we have $(U(\underline{\ell}))_{\mathbf{Z}} = U^{!}_{\mathbf{Z}}(\underline{\ell})$. Let $\Delta^{(n)}: U(\underline{\ell}) \to (End\, L_1) \otimes \cdots \otimes (End\, L_n)$ be the usual diagonal map from the universal enveloping algebra into the tensor product of n associative algebras $(End\, L_i)$ (each L_i an $\underline{\ell}$-module).

Proposition 4.6.4 *If A_i is an additive subgroup of the module L_i for $i = 1, 2, \ldots, n$ and each A_i is preserved by the monomials in (3.1.12)-(3.1.14), then the tensor product $A_1 \otimes \cdots \otimes A_n$ is preserved by the image of the monomials in (3.1.12)-(3.1.14) under the diagonal map $\Delta^{(n)}$.*

Proof: To see this is true, it is enough to check the result holds for monomials of the form $\frac{\iota(e_\alpha)^m}{m!}$, $\alpha \in \Delta_R(\underline{g})$, and of the form $\Lambda_s(X_{n,i})$. The first monomial acts as

$$\Delta^{(n)}\left(\frac{\iota(e_\alpha)^m}{m!}\right) = \sum_{\substack{j_1+j_2+\cdots+j_n=m \\ j_k\geq 0}} \frac{\iota(e_\alpha)^{j_1}}{j_1!} \otimes \cdots \otimes \frac{\iota(e_\alpha)^{j_n}}{j_n!} \tag{4.6.6}$$

on the module $L_1 \otimes \cdots \otimes L_n$. Since each $\frac{\iota(e_\alpha)^{j_k}}{j_k!}$ preserves each tensorand A_i, the expression in (4.6.6) preserves the tensor product $A_1 \otimes \cdots \otimes A_n$.

Using (4.3.4) and Proposition 4.3.5 we see that the monomial $\Lambda_s(X_{n,i})$ acts on the tensor product module as

$$\begin{aligned}
&\Delta^{(n)}\left(\Lambda_s(X_{n,i})\right) = \\
&= \Delta^{(n)}\left(\sum_{|\lambda|=s}\left(\prod_{j\geq 1}\frac{\alpha_i(-nj)^{m_j(\lambda)}}{j^{m_j(\lambda)}m_j(\lambda)!}\right)\right) \\
&= \sum_{|\lambda|=s}\prod_{j\geq 1}\left(\sum_{\substack{l_1+\cdots+l_n=m_j(\lambda) \\ l_k\geq 0}}\frac{1}{l_1!}\left(\frac{\alpha_i(-nj)}{j}\right)^{l_1}\otimes\cdots\otimes\frac{1}{l_n!}\left(\frac{\alpha_i(-nj)}{j}\right)^{l_n}\right) \\
&= \sum_{\substack{l_1+\cdots+l_n=s \\ l_k\geq 0}}\left(\sum_{|\lambda|=l_1}\prod_{j\geq 1}\frac{\alpha_i(-nj)^{m_j(\lambda)}}{j^{m_j(\lambda)}m_j(\lambda)!}\right)\otimes\cdots\otimes\left(\sum_{|\lambda|=l_n}\prod_{j\geq 1}\frac{\alpha_i(-nj)^{m_j(\lambda)}}{j^{m_j(\lambda)}m_j(\lambda)!}\right) \\
&= \sum_{\substack{l_1+\cdots+l_n=s \\ l_k\geq 0}}\Lambda_{l_1}(X_{n,i})\otimes\cdots\otimes\Lambda_{l_n}(X_{n,i}).
\end{aligned}$$

Since each $\Lambda_{l_k}(X_{n,i})$ preserves each A_i, then the image of $\Lambda_s(X_{n,i})$ under $\Delta^{(n)}$ also preserves $A_1 \otimes \cdots \otimes A_n$. □

Theorem 4.6.5 $U^!_{\mathbf{Z}}(\underline{\ell}) = (U(\underline{\ell}))_{\mathbf{Z}}$.

Proof: By Theorem 4.6.3 we only need to show $(U(\underline{\ell}))_{\mathbf{Z}} \subset U^!_{\mathbf{Z}}(\underline{\ell})$. Since the monomials in (3.1.12)-(3.1.14) from a $\mathbf{C}$-basis for $U(\underline{\ell})$, we know that any element of $(U(\underline{\ell}))_{\mathbf{Z}}$ can bc written as a $\mathbf{C}$-linear combination of the (ordered) monomials in (3.1.12)-(3.1.14).

Let $u \in (U(\underline{\ell}))_{\mathbf{Z}}$ and suppose u is given by

$$u = \sum_i c_i f_i \tag{4.6.7}$$

where each f_i is an ordered product of the monomials in (3.1.12)-(3.1.14), $c_i \in \mathbf{C}$, and only finitely many of the $c_i f_i$ are nonzero. Since $u \in U(\underline{\ell})_{\mathbf{Z}}$, u must preserve the integral form of each $M(\alpha)$ for $\alpha \in L''$. If u is to preserve the $M(\alpha)_{\mathbf{Z}}$'s, we must check that each summand of degree γ (cf. Remark 4.6.1), i.e.,

$$\sum_{\substack{j \\ \deg f_j = \gamma}} c_j f_j,$$

in (4.6.7) preserves the $M(\alpha)_{\mathbf{Z}}$'s.

Now consider the summand of degree γ in the expression (4.6.7) for u. Recall the filtration of $U(\underline{\ell})$ used in §3.1, i.e., the "usual" filtration. Let f_{j_1} be the ordered monomial of largest "filtration" degree. Thus we may write f_{j_1} as the ordered monomial

$$\begin{aligned} c_{j_1} \frac{\iota(e_{-\alpha_{k_1}})^{a_{k_1}}}{(a_{k_1})!} \cdots \frac{\iota(e_{-\alpha_{k_r}})^{a_{k_r}}}{(a_{k_r})!} \cdot \Lambda_{l_1}(X_{n_1,i_1}) \cdots \\ \cdots \Lambda_{l_{s-1}}(X_{n_{s-1},i_{s-1}}) \cdot \Lambda_{l_s}(X_{0,0}) \cdot \frac{\iota(e_{\alpha_{m_1}})^{b_{m_1}}}{(b_{m_1})!} \cdots \frac{\iota(e_{\alpha_{m_t}})^{b_{m_t}}}{(b_{m_t})!}, \end{aligned} \tag{4.6.8}$$

where the degree of $f_{j_1} = J = a_{k_1} + \cdots + a_{k_r} + l_1 + \cdots + l_s + b_{m_1} + \cdots + b_{m_t}$, α_{k_i}, $\alpha_{m_i} \in \Delta_{\mathbf{R}}(\underline{\ell})$, and $a_{k_i}, l_i, b_{m_i}, |n_i| \in \mathbf{N}$.

Next consider the $\underline{\ell}$-module

$$\mathsf{M} = \underbrace{M(\alpha_{k_1}) \otimes \cdots \otimes M(\alpha_{k_1})}_{a_{k_1} \text{ times}} \otimes \cdots$$

$$\cdots \otimes \underbrace{M(\alpha_{k_r}) \otimes \cdots \otimes M(\alpha_{k_r})}_{a_{k_r} \text{ times}} \otimes \underbrace{M(\Lambda_{i_1}) \otimes \cdots \otimes M(\Lambda_{i_1})}_{l_1 \text{ times}} \otimes \cdots$$

$$\cdots \otimes \underbrace{M(\Lambda_{i_{s-1}}) \otimes \cdots \otimes M(\Lambda_{i_{s-1}})}_{l_{s-1} \text{ times}} \otimes \underbrace{M(\Lambda_0) \otimes \cdots \otimes M(\Lambda_0)}_{l_s \text{ times}} \otimes$$

$$\otimes \underbrace{M(-\alpha_{m_1}) \otimes \cdots \otimes M(-\alpha_{m_1})}_{b_{m_1} \text{ times}} \otimes \cdots \otimes \underbrace{M(-\alpha_{m_t}) \otimes \cdots \otimes M(-\alpha_{m_t})}_{b_{m_t} \text{ times}}$$

where the Λ_{i_k} are fundamental weights, i.e., $\langle \Lambda_{i_k}, \alpha_{i_j} \rangle = \delta_{k,j}$.

By Proposition 4.6.4, if the summand of degree γ preserves all the $M(\alpha)_{\mathbf{Z}}$'s, then $\Delta^{(J)}\left(\sum_j c_j f_j\right)$ will preserve

$$\mathsf{M}_{\mathbf{Z}} = \underbrace{M(\alpha_{k_1})_{\mathbf{Z}} \otimes \cdots M(\alpha_{k_1})_{\mathbf{Z}}}_{a_{k_1} \text{ times}} \otimes \cdots \otimes \underbrace{M(-\alpha_{m_t})_{\mathbf{Z}} \otimes \cdots \otimes M(-\alpha_{m_t})_{\mathbf{Z}}}_{b_{m_t} \text{ times}} .$$

The vector

$$v^{\otimes} = \underbrace{\overline{\iota(e_{\alpha_{k_1}})} \otimes \cdots \otimes \overline{\iota(e_{\alpha_{k_1}})}}_{a_{k_1} \text{ times}} \otimes \cdots \otimes \underbrace{\overline{\iota(e_{-\alpha_{m_t}})} \otimes \cdots \otimes \overline{\iota(e_{-\alpha_{m_t}})}}_{b_{m_t} \text{ times}}$$

is an element of $\mathsf{M}_{\mathbf{Z}}$ and is contained in $\sum_{\mu_1,\ldots,\mu_J} M(\alpha_{k_1})_{\mu_1} \otimes \cdots \otimes M(-\alpha_{m_t})_{\mu_J}$. Then

$$\begin{aligned} \Delta^{(J)}\left(\sum_j c_j f_j\right) \cdot v^{\otimes} \in \; & \pm c_{j_1} \overline{\alpha_{k_1}(-1)} \otimes \cdots \otimes \overline{\alpha_{k_r}(-1)} \otimes \overline{\iota(1)} \otimes \cdots \\ & \cdots \otimes \overline{\iota(1)} \otimes \overline{\alpha_{m_1}(-1)} \otimes \cdots \otimes \overline{\alpha_{m_t}(-1)} \\ & + \sum_{\substack{\mu_1,\ldots,\mu_J \\ \text{not all } 0}} M(\alpha_{k_1})_{\mu_1} \otimes \cdots \otimes M(-\alpha_{m_t})_{\mu_J}. \end{aligned}$$

For $\Delta^j(\sum_j c_j f_j)$ to remain in $\mathsf{M}_{\mathbf{Z}}$, the leading term must be in $\mathsf{M}_{\mathbf{Z}}$. Thus $c_{j_i} \in \mathbf{Z}$. To see the other c_j's are integers, repeat the above argument using the element $u - c_{j_1} f_{j_1} \in (U(\underline{\ell}))_{\mathbf{Z}}$. Hence each $u \in (U(\underline{\ell}))_{\mathbf{Z}}$ is a $\mathbf{Z}$-linear combination of the monomials in (3.1.12)-(3.1.14), and therefore $(U(\underline{\ell}))_{\mathbf{Z}} \subset U^{\sharp}_{\mathbf{Z}}(\underline{\ell})$. □

4.7 A description for the unequal root length affine Lie algebras

In this section we turn our attention to the unequal root length affines introduced in Sections 2.2 through 2.6.

First we consider the subalgebras $\underline{\ell}_{[0]}$ of $\underline{\ell}$ where $\underline{\ell}_{[0]}$ is of type $C_n^{(1)}$, $B_n^{(1)}$, $F_4^{(1)}$, or $G_n^{(1)}$, respectively, and $\underline{\ell}$ is of type $A_{2n-1}^{(1)}$, $D_{n+1}^{(1)}$, $E_6^{(1)}$, or $D_4^{(1)}$, respectively. The subalgebra $\underline{\ell}_{[0]}{}'$ is mapped to the Lie algebra P^1/DP^0 using (4.5.20):

$$\begin{aligned} e_i &\longmapsto \iota(e_{\alpha_i})^+_{[0]} \\ f_i &\longmapsto \epsilon(\alpha_i,-\alpha_i)\iota(e_{-\alpha_i})^+_{[0]} \\ h_i &\longmapsto \alpha^+_{i,[0]}(-1) \end{aligned} \tag{4.7.1}$$

(cf. Proposition 4.5.5). We may also view any module M for the algebra $\underline{\ell}$ as an $\underline{\ell}_{[0]}$-module by restriction. So in particular, each of the $\underline{\ell}$-modules $M(\alpha)$ for $\alpha \in L''$ is an $\underline{\ell}_{[0]}$-module.

As in §4.6, the subset $\underline{\ell}_{[0],\mathbf{Z}}$ of $\underline{\ell}_{[0]}$ given by

$$\underline{\ell}_{[0]} \cap (V_{L'})_{\mathbf{Z}}/DV_{L'\mathbf{Z}}, \tag{4.7.2}$$

is a $\mathbf{Z}$-form of $\underline{\ell}_{[0]}$. Since $\underline{\ell}_{[0],\mathbf{Z}} \subset \underline{\ell}_{\mathbf{Z}}$, $\underline{\ell}_{[0],\mathbf{Z}}$ preserves all of the $M(\alpha)_{\mathbf{Z}}$'s.

Let $U^{!}_{\mathbf{Z}}(\underline{\ell}_{[0]})$ denote the $\mathbf{Z}$-subalgebra described in Theorem 4.2.7 with the adjustment on the degree operator discussed in § 4.6.

Theorem 4.7.1 *Let $\left(U(\underline{\ell}_{[0]})\right)_{\mathbf{Z}}$ denote the* $\mathbf{Z}$*-subalgebra of $U(\underline{\ell}_{[0]})$ which preserves all the modules $M(\alpha)_{\mathbf{Z}}$ for $\alpha \in L''$. Then $\left(U(\underline{\ell}_{[0]})\right)_{\mathbf{Z}} = U^{!}_{\mathbf{Z}}(\underline{\ell}_{[0]})$.*

Proof: As in Theorem 4.6.3 we first show that $\left(U(\underline{\ell}_{[0]})\right)_{\mathbf{Z}}$ contains $\frac{(e_i)^n}{n!}$ and $\frac{(f_i)^n}{n!}$ for e_i and f_i in (4.7.1). If the graph automorphism ν is an involution, e_i is given by

$$\iota(e_{\alpha_i})^+_{[0]} = \begin{cases} \iota(e_{\alpha_i}) & \text{if } \ \nu\alpha_i = \alpha_i \\ \iota(e_{\alpha_i}) + \iota(e_{\nu\alpha_i}) & \text{if } \ \nu\alpha_i \neq \alpha_i. \end{cases}$$

Recall that ν is a nontrivial graph automorphism of the Dynkin diagram associated to the root system of A_{2n-1}, D_{n+1} or E_6, we have $\langle \alpha_i, \nu\alpha_i \rangle \geq 0$ for each $\alpha_i \in \Pi^+$. Hence we obtain

$$\frac{\left(\iota(e_{\alpha_i})^+_{[0]}\right)^n}{n!} = \begin{cases} \frac{\left(\iota(e_{\alpha_i})\right)^n}{n!} & \text{if } \ \nu\alpha_i = \alpha_i \\ \sum_{j=0}^n \frac{\left(\iota(e_{\alpha_i})\right)^{n-j}}{(n-j)!}\frac{\left(\iota(e_{\nu\alpha_i})\right)^j}{j!} & \text{if } \ \nu\alpha_i \neq \alpha_i \end{cases} \tag{4.7.3}$$

inside the larger simply-laced algebra $\underline{\ell}$. Thus by Theorem 4.6.3 $\frac{(e_i)^n}{n!}$ and $\frac{(f_i)^n}{n!}$ are elements of $\left(U(\underline{\ell}_{[0]})\right)_{\mathbf{Z}}$.

If ν is the graph automorphism of D_4 given in §2.4, e_i is given by

$$\iota(e_{\alpha_i})^+_{[0]} = \begin{cases} \iota(e_{\alpha_2}) & \text{if } i = 0,\ 2 \\ \iota(e_{\alpha_1}) + \iota(e_{\alpha_3}) + \iota(e_{\alpha_4}) & \text{if } i = j. \end{cases}$$

By Lemma 5.1 $\langle \nu\alpha, \alpha\rangle = \langle \nu^2\alpha, \alpha\rangle = 0$ for $\alpha \in \Delta$, so we obtain

$$\frac{\left(\iota(e_{\alpha_i})^+_{[0]}\right)^n}{n!} = \begin{cases} \frac{(\iota(e_{\alpha_2}))^n}{n!} & \text{if } i = 0,\ 2 \\ \sum_{j=0}^{n}\sum_{k=0}^{n-j} \frac{(\iota(e_{\alpha_i}))^{n-j-k}}{(n-j-k)!}\frac{(\iota(e_{\nu\alpha_i}))^j}{j!}\frac{\left(\iota(e_{\nu^2\alpha_i})\right)^k}{k!} & \text{if } i = j \end{cases} \tag{4.7.4}$$

inside the algebra $\underline{\ell}$. Again Theorem 4.6.3 implies $\frac{(e_i)^n}{n!}$ and $\frac{(f_i)^n}{n!}$ are elements of $\left(U(\underline{\ell}_{[0]})\right)_{\mathbf{Z}}$. The remaining argument to see $U^!_{\mathbf{Z}}(\underline{\ell}_{[0]}) \subset \left(U(\underline{\ell}_{[0]})\right)_{\mathbf{Z}}$ is similar to that found in Theorem 4.6.3.

For the reverse inclusion we again use the work done in the larger simply-laced algebra. As before, any element $u \in \left(U(\underline{\ell}_{[0]})\right)_{\mathbf{Z}}$ can be written as a $\mathbf{C}$-linear combination of the monomials found in (3.1.12)-(3.1.14). If this expression for u contains coefficients other than integers, we would have a contradiction to Theorem 4.6.5: this expression could be "expanded" to an expression in $U(\underline{\ell})$ which is not a $\mathbf{Z}$-linear combination of monomials, i.e., $U^!_{\mathbf{Z}}(\underline{\ell}) \neq (U(\underline{\ell}))_{\mathbf{Z}}$. Hence $\left(U(\underline{\ell}_{[0]})\right) \subset U^!_{\mathbf{Z}}(\underline{\ell}_{[0]})$ and $\left(U(\underline{\ell}_{[0]})\right)_{\mathbf{Z}} = U^!_{\mathbf{Z}}(\underline{\ell}_{[0]})$. □

Next consider the subalgebras $\underline{\ell}^{(\tau)}$ of $\underline{\ell}$ where $\underline{\ell}^{(\tau)}$ is of type $A^{(2)}_{2n-1}$, $D^{(2)}_{n+1}$ or $E^{(2)}_6$, respectively, and $\underline{\ell}$ is of type A_{2n-1}, D_{n+1} or E_6, respectively.

The subalgebra $\underline{\ell}^{(\tau)}$ is mapped to the Lie algebra P^1/DP^0 via the map (cf. (4.7.1))

$$
\begin{aligned}
e_i &\longmapsto \iota(e_{\alpha_i})^+_{[0]} \\
f_i &\longmapsto \epsilon(\alpha_i, -\alpha_i)\iota(e_{-\alpha_i})^+_{[0]} \\
h_i &\longmapsto \alpha^+_{i,[0]}(-1) \\
e_0 &\longmapsto \iota(e_\gamma)^+_{[1]} \\
f_0 &\longmapsto \epsilon(\gamma, -\gamma)\iota(e_{-\gamma})^+_{[1]} \\
h_0 &\longmapsto \gamma^+_{[0]}(-1)
\end{aligned}
\tag{4.7.5}
$$

where $i \in I$ and γ is as in [K-K-L-W].

As for the subalgebras $\underline{\ell}_{[0]}$, we may view any $\underline{\ell}$-module M as an $\underline{\ell}^{(\tau)}$-module via restriction of the $\underline{\ell}$ action to the subalgebra $\underline{\ell}^{(\tau)}$. Thus the modules $M(\alpha)$ for $\alpha \in L''$ are $\underline{\ell}^{(\tau)}$-modules.

Let $\underline{\ell}^{(\tau)}{}_{\mathbf{Z}}$ be the subset of $\underline{\ell}^{(\tau)}$ which is represented by elements in $(V_{L'})_{\mathbf{Z}}$. Thus $\underline{\ell}^{(\tau)}{}_{\mathbf{Z}}$ is given by $\underline{\ell}^{(\tau)} \cap (V_{L'})_{\mathbf{Z}}/D\,(V_{L'})_{\mathbf{Z}}$, and is a $\mathbf{Z}$-form of $\underline{\ell}^{(\tau)}$ since the Chevalley basis of $\left(\underline{\ell}^{(\tau)}\right)'$ (the derived subalgebra) along with $r \cdot (-L(0) + \frac{1}{24} dim\, \underline{\mathrm{h}}\, \iota(1)_{-1})$ are elements of $\underline{\ell}^{(\tau)}{}_{\mathbf{Z}}$ (cf. (4.6.3)).

Let $U^\sharp_{\mathbf{Z}}(\underline{\ell}^{(\tau)})$ denote the $\mathbf{Z}$-subalgebra described in Theorem 4.2.6 with the degree operator replaced by an appropriate integral multiple of itself (cf. (4.6.3)).

Theorem 4.7.2 *Let $\left(U(\underline{\ell}^{(\tau)})\right)_{\mathbf{Z}}$ denote the $\mathbf{Z}$-subalgebra of $U(\underline{\ell}^{(\tau)})$ which preserves all the modules $M(\alpha)$ for $\alpha \in L''$. Then $\left(U(\underline{\ell}^{(\tau)})\right)_{\mathbf{Z}} = U^\sharp_{\mathbf{Z}}(\underline{\ell}^{(\tau)})$.*

Proof: The graph automorphism ν used to define $\underline{\ell}^{(\tau)}$ is the same as in the $\underline{\ell}_{[0]}$ case. Hence a similar argument to the one found in the proof of Theorem 4.7.1 gives the desired result. □

The affine algebras $\underline{\ell}^{(\tau)}$ of type $A^{(2)}_{2n}$ must be dealt with in a slightly different manner. Recall the Chevalley basis of $\underline{\ell}^{(\tau)}$ contains elements of the form $\overline{\iota(e_\alpha)}^+_{[n]}$ for $\alpha \in \Delta_R(\underline{\ell}^{(\tau)})$. If $\alpha \in \Delta_{-1}$ then $\overline{\iota(e_\alpha)}^+_{[n]} = \sqrt{2}\iota(e_\alpha)^+_{[n]}$, and so $\overline{\iota(e_\alpha)}^+_{[n]}$ would not be represented in $(V_{L'})_{\mathbf{Z}}/D\,(V_{L'})_{\mathbf{Z}}$. To get around this difficulty, we must go back to the description of $(V_{L'})_{\mathbf{Z}}$. Instead of the subalgebra $(V_{L'})_{\mathbf{Z}}$ used in Theorem 4.4.1, consider the smallest $\mathbf{Z}$-subring of $V_{L'}$, $\left(\overline{V_{L'}}\right)_{\mathbf{Z}}$, containing all the $\overline{\iota(a)}$ for $a \in \hat{L}$, and closed under action by the operators $\frac{(L(-1))^n}{n!}$, $n \in \mathbf{N}$. As in Theorem 4.4.1, $\left(\overline{V_{L'}}\right)_{\mathbf{Z}}$ is similarly shown to be an integral form for $V_{L'}$. Then replace $(V_{L'})_{\mathbf{Z}}$ with $\left(\overline{V_{L'}}\right)_{\mathbf{Z}}$ in (4.6.1).

Let $U^{!}_{\mathbf{Z}}(\underline{\ell}^{(\tau)})$ denote the $\mathbf{Z}$-subalgebra described in Theorem 4.2.7 with the degree operator replaced by an appropriate integral multiple of itself (cf. (4.6.3)).

Theorem 4.7.3 *Let $\left(U(\underline{\ell}^{(\tau)})\right)_{\mathbf{Z}}$ denote the $\mathbf{Z}$-subalgebra of $U(\underline{\ell}^{(\tau)})$ which preserves all the modules $M(\alpha)$ for $\alpha \in L''$. Then $\left(U(\underline{\ell}^{(\tau)})\right)_{\mathbf{Z}} = U^{!}_{\mathbf{Z}}(\underline{\ell}^{(\tau)})$.*

Proof: If $\alpha_i \in \Pi_1$ (cf. §2.6), then an argument similar to that found in the proof of Theorem 4.7.2 shows $\frac{(e_i)^n}{n!}$ and $\frac{(f_i)^n}{n!}$ are elements of $\left(U(\underline{\ell}^{(\tau)})\right)_{\mathbf{Z}}$. If $\alpha_i \in \Pi_{-1}$ (or equivalently, $i = n$), then

$$\begin{aligned}
[\overline{\iota(e_{\alpha_n})}, \overline{\iota(e_{\nu\alpha_n})}] &= 2\epsilon(\alpha_n, \nu\alpha_n)\overline{\iota(e_{\alpha_n+\nu\alpha_n})} \\
&= 2\epsilon(\alpha_n, \alpha_{n+1})\overline{\iota(e_{\alpha_n+\alpha_{n+1}})}.
\end{aligned}$$

A straightforward calculation shows

$$\begin{aligned}
\frac{\left(\overline{\iota(e_{\alpha_n})}^{+}_{[0]}\right)^k}{k!} &= \sum_{j=0}^{k} \frac{\left(\overline{\iota(e_{\alpha_n})}\right)^{k-j}}{(k-j)!} \frac{\left(\overline{\iota(e_{\alpha_{n+1}})}\right)^{j}}{j!} \\
&+ \sum_{j=0}^{k-2} \frac{\left(\overline{\iota(e_{\alpha_n})}\right)^{k-j-2}}{(k-j-2)!} \frac{\left(\overline{\iota(e_{\alpha_{n+1}})}\right)^{j}}{j!} \left(\epsilon(\alpha_n, \alpha_{n+1})\overline{\iota(e_{\alpha_n+\alpha_{n+1}})}\right) \\
&\vdots \\
&+ \sum_{j=0}^{[\frac{k}{2}]} \frac{\left(\overline{\iota(e_{\alpha_n})}\right)^{[\frac{k}{2}]-j}}{\left([\frac{k}{2}]-j\right)!} \frac{\left(\overline{\iota(e_{\alpha_{n+1}})}\right)^{j}}{j!} \frac{\left(\epsilon(\alpha_n, \alpha_{n+1})\overline{\iota(e_{\alpha_n+\alpha_{n+1}})}\right)^{k-2}}{(k-2)!}.
\end{aligned}$$

Now apply an argument similar to that found in Theorem 4.7.2 to complete the proof.

□

Bibliography

[B] Borcherds, R. E. Proc. Natl. Acad. Sci. USA **83**, 3068-3071 (1986).

[Chev] Chevalley, C. *Sur certain groupes simples.* Tôhoku Math. J. (2) **7**, 14-66 (1955).

[F-K] Frenkel, I. B. and Kac, V. G. *Basic representations of affine Lie algebras and dual resonance models.* Invent. Math. **62**, 23-66 (1980).

[F-L-M] Frenkel, I. B., Lepowsky, J., and Meurman, A. *Vertex Operator Algebras and the Monster.* Pure and Applied Math., Academic Press, 1988.

[G] Garland, H. *The arithmetic theory of loop algebras.* J. Algebra. **53**, 490-551 (1978).

[H] Humphreys, J. E. *Introduction to Lie Algebras and Representation Theory.* Springer-Verlag, New York, 1972.

[Kac] Kac, V. G. *Infinite dimesional Lie algebras.* Birkhäuser, Boston, 1983.

[K-K-L-W] Kac, V. G., Kazhdan, D. A., Lepowsky, J. and Wilson, R. *Realization of the basic representations of the Euclidean Lie algebras.* Advances in Math. **42**, 83-112 (1981).

[Kost] Kostant, B. *Groups over* **Z**, in *Algebraic Groups and Discontinuous Subgroups*, Proceedings of Symposia in Pure Mathematics. **109**, 90-98, American Math. Soc., Providence, RI (1966).

[L] Lepowsky, J. *Lectures on Kac-Moody Lie algebras.* Université Paris IV, Spring 1978.

[L1] Lepowsky, J. *Calculus of twisted vertex operators.* Proc. of Nat. Acad. Sci., U.S.A., **82**, 8295-8299 (1985).

[L-P] Lepowsky, J. and Primc, M. Contemporary Math. **46**, Amer. Math. Soc. (1985).

[L-W1] Lepowsky, J. and Wilson, R. L. *Construction of the affine Lie algebra* $A_1^{(1)}$. Commun. Math. Physics. **62**, 43-53 (1978).

[L-W2] Lepowsky, J. and Wilson, R. L. *The structure of standard modules, I: Universal algebras and the Rogers-Ramanujan identities.* Invent. Math. **77**, 199-290 (1984).

[Mac] Macdonald, I. G. *Symmetric functions and Hall polynomials.* Clarendon Press, Oxford, 1979.

[M] Mitzman, D. *Integral bases for affine Lie algebras and their universal enveloping algebras.* Contemporary Math. **40**, 1985.

[Seg] Segal, G. *Unitary representations of some infinite-dimensional groups.* Commun. Math. Phy. **80**, 301-342 (1981).

[Stein] Steinberg, R. *Lectures on Chevalley Groups.* Yale University mimeographed notes, (1967).

[T] Tits, J. *Théorie des groupes.* Résumé de cours, Annuaire du Collège de France, 75-87 (1980-81).

MEMOIRS of the American Mathematical Society

SUBMISSION. This journal is designed particularly for long research papers (and groups of cognate papers) in pure and applied mathematics. The papers, in general, are longer than those in the TRANSACTIONS of the American Mathematical Society, with which it shares an editorial committee. Mathematical papers intended for publication in the Memoirs should be addressed to one of the editors:

Ordinary differential equations, partial differential equations and applied mathematics to ROGER D. NUSSBAUM, Department of Mathematics, Rutgers University, New Brunswick, NJ 08903

Harmonic analysis, representation theory and Lie theory to AVNER D. ASH, Department of Mathematics, The Ohio State University, 231 West 18th Avenue, Columbus, OH 43210

Abstract analysis to MASAMICHI TAKESAKI, Department of Mathematics, University of California, Los Angeles, CA 90024

Real and harmonic analysis to DAVID JERISON, Department of Mathematics, M.I.T., Rm 2–180, Cambridge, MA 02139

Algebra and algebraic geometry to JUDITH D. SALLY, Department of Mathematics, Northwestern University, Evanston, IL 60208

Geometric topology and general topology to JAMES W. CANNON, Department of Mathematics, Brigham Young University, Provo, UT 84602

Algebraic topology and differential topology to RALPH COHEN, Department of Mathematics, Stanford University, Stanford, CA 94305

Global analysis and differential geometry to JERRY L. KAZDAN, Department of Mathematics, University of Pennsylvania, E1, Philadelphia, PA 19104-6395

Probability and statistics to RICHARD DURRETT, Department of Mathematics, Cornell University, Ithaca, NY 14853-7901

Combinatorics and number theory to CARL POMERANCE, Department of Mathematics, University of Georgia, Athens, GA 30602

Logic, set theory, general topology and universal algebra to JAMES E. BAUMGARTNER, Department of Mathematics, Dartmouth College, Hanover, NH 03755

Algebraic number theory, analytic number theory and modular forms to AUDREY TERRAS, Department of Mathematics, University of California at San Diego, La Jolla, CA 92093

Complex analysis and nonlinear partial differential equations to SUN-YUNG A. CHANG, Department of Mathematics, University of California at Los Angeles, Los Angeles, CA 90024

All other communications to the editors should be addressed to the Managing Editor, DAVID J. SALTMAN, Department of Mathematics, University of Texas at Austin, Austin, TX 78713.

General instructions to authors for

PREPARING REPRODUCTION COPY FOR MEMOIRS

For more detailed instructions send for AMS booklet, "A Guide for Authors of Memoirs." Write to Editorial Offices, American Mathematical Society, P.O. Box 6248, Providence, R.I. 02940-6248.

MEMOIRS are printed by photo-offset from camera copy fully prepared by the author. This means that the finished book will look exactly like the copy submitted. Thus the author will want to use a good quality typewriter with a new, medium-inked black ribbon, and submit clean copy on the appropriate model paper.

Model Paper, provided at no cost by the AMS, is paper marked with blue lines that confine the copy to the appropriate size.

Special Characters may be filled in carefully freehand, using dense black ink, or **INSTANT** ("rub-on") **LETTERING** may be used. These may be available at a local art supply store.

Diagrams may be drawn in black ink either directly on the model sheet, or on a separate sheet and pasted with rubber cement into spaces left for them in the text. Ballpoint pen is not acceptable.

Page Headings (Running Heads) should be centered, in CAPITAL LETTERS (preferably), at the top of the page — just above the blue line and touching it.

LEFT-hand, EVEN-numbered pages should be headed with the AUTHOR'S NAME;

RIGHT-hand, ODD-numbered pages should be headed with the TITLE of the paper (in shortened form if necessary).

Exceptions: PAGE 1 and any other page that carries a display title require NO RUNNING HEADS.

Page Numbers should be at the top of the page, on the same line with the running heads.

LEFT-hand, EVEN numbers — flush with left margin;

RIGHT-hand, ODD numbers — flush with right margin.

Exceptions: PAGE 1 and any other page that carries a display title should have page number, centered below the text, on blue line provided.

FRONT MATTER PAGES should be numbered with Roman numerals (lower case), positioned below text in same manner as described above.

MEMOIRS FORMAT

It is suggested that the material be arranged in pages as indicated below.
Note: Starred items (*) are requirements of publication.

Front Matter (first pages in book, preceding main body of text).

Page i — *Title, *Author's name.

Page iii — Table of contents.

Page iv — *Abstract (at least 1 sentence and at most 300 words).

Key words and phrases, if desired. (A list which covers the content of the paper adequately enough to be useful for an information retrieval system.)

*<u>*1991 Mathematics Subject Classification*</u>. This classification represents the primary and secondary subjects of the paper, and the scheme can be found in Annual Subject Indexes of MATHEMATICAL REVIEWS beginnning in 1990.

Page 1 — Preface, introduction, or any other matter not belonging in body of text.

Footnotes: *Received by the editor date.
Support information — grants, credits, etc.

First Page Following Introduction – Chapter Title (dropped 1 inch from top line, and centered). Beginning of Text.

Last Page (at bottom) – Author's affiliation.

Recent Titles in This Series

(Continued from the front of this publication)

431 **Thomas G. Goodwillie,** A multiple disjunction lemma for smooth concordance embeddings

430 **G. M. Benkart, D. J. Britten, and F. W. Lemire,** Stability in modules for classical Lie algebras: A constructive approach

429 **Etsuko Bannai,** Positive definite unimodular lattices with trivial automorphism groups

428 **Loren N. Argabright and Jesús Gil de Lamadrid,** Almost periodic measures

427 **Tim D. Cochran,** Derivatives of links: Milnor's concordance invariants and Massey's products

426 **Jan Mycielski, Pavel Pudlák, and Alan S. Stern,** A lattice of chapters of mathematics: Interpretations between theorems

425 **Wiesław Pawłucki,** Points de Nash des ensembles sous-analytiques

424 **Yasuo Watatani,** Index for C^*-subalgebras

423 **Shek-Tung Wong,** The meromorphic continuation and functional equations of cuspidal Eisenstein series for maximal cuspidal subgroups

422 **A. Hinkkanen,** The structure of certain quasisymmetric groups

421 **H. W. Broer, G. B. Huitema, F. Takens, and B. L. J. Braaksma,** Unfoldings and bifurcations of quasi-periodic tori

420 **Tom Lindstrøm,** Brownian motion on nested fractals

419 **Ronald S. Irving,** A filtered category $\mathcal{O}_S$ and applications

418 **Hiroaki Hijikata, Arnold K. Pizer, and Thomas R. Shemanske,** The basis problem for modular forms on $\Gamma_0(N)$

417 **John B. Conway, Domingo A. Herrero, and Bernard B. Morrel,** Completing the Riesz-Dunford functional calculus

416 **Jacek Nikiel,** Topologies on pseudo-trees and applications

415 **Amnon Neeman,** Ueda theory: Theorems and problems

414 **Nikolaus Vonessen,** Actions of linearly reductive groups on affine PI-algebras

413 **Francesco Brenti,** Unimodal log-concave and polya frequency sequences in combinatorics

412 **T. Levasseur and J. T. Stafford,** Rings of differential operators on classical rings of invariants

411 **Matthew G. Brin and** [redacted]-movable

410 **P. L.** [redacted]s and geome[redacted]

409 **William Arveson,** Continuous analogues of Frock space

408 **Idun Reiten and Michel Van den Bergh,** Two-dimensional tame and maximal orders of finite representation type

407 **Fokko du Cloux,** Représentations de longueur finie des groupes de Lie résolubles

406 **H. Toruńczyk and J. West,** Fibrations and bundles with Hilbert cube manifold fibers

405 **J. N. Brooks and J. B. Strantzen,** Blaschke's rolling theorem in $\mathbb{R}^n$

404 **Andreas Blass and Andre Scedrov,** Freyd's models for the independence of the axiom of choice

403 **Tomasz Przebinda,** The oscillator duality correspondence for the pair $O(2,2), SP(2,\mathbb{R})$

402 **Hans-Otto Walther,** Hyperbolic periodic solutions, heteroclinic connections and transversal homoclinic points in autonomous differential delay equations

401 **Roger D. Nussbaum,** Iterated nonlinear maps and Hilbert's projective metric, II

400 **Nikolaos Mandouvalos,** Scattering operator, Eisenstein series, inner product formula and "Maass-Selberg" relations for Kleinian groups

(See the AMS catalogue for earlier titles)